Exile Air

Spirit of Little Norway

Exile Air

World War II's "Little Norway" in Toronto and Muskoka

Andrea Baston

Photography and photographic editing by

Candis Jones

Old Stone Books Ltd.
Toronto, 2017

FRONT COVER

Pilot trainees with Curtiss P-36 aircraft at the Little Norway training centre, Toronto Island, Toronto, Ontario. PHOTO: LIBRARY AND ARCHIVES CANADA / DEPARTMENT OF NATIONAL DEFENCE FONDS / PA-136047.

BACK COVER

Sign welcomes visitors to the Little Norway Memorial and Muskoka Air Terminal, located at Muskoka Airport, Gravenhurst, Ontario. PHOTOGRAPHER: CANDIS JONES.

FACING TITLE PAGE

Royal Norwegian Air Force pilots board a *Spirit of Little Norway* Fairchild training airplane. This is one of three *Spirit of Little Norway* aircraft which were purchased over the war with funds voluntarily donated by Little Norway's personnel. PHOTO: LITTLE NORWAY MEMORIAL.

FACING TABLE OF CONTENTS

Norwegian airman returns from swimming in Long (Dotty) Lake at Vesle Skaugum, near Huntsville, Ontario, Canada. The Royal Norwegian Air Force used the old fur trading post as a training and recreation centre during World War II. PHOTO: GRAVENHURST ARCHIVES.

Old Stone Books Ltd.
Tel. (647) 965-5397
email: info@oldstonebooks.com
website: oldstonebooks.com

Library and Archives Canada Cataloguing in Publication

Baston, Andrea, 1950–, author
Exile air : World War II's "Little Norway" in Toronto and Muskoka / Andrea Baston ; photography and photographic editing by Candis Jones.

Includes bibliographical references and index.
ISBN 978-0-9920903-2-6 (softcover)

1. Norway. Luftforsvaret—History—20th century. 2. British Commonwealth Air Training Plan. 3. World War, 1939–1945—Aerial operations, Norwegian. 4. World War, 1939–1945—Participation, Norwegian. 5. Flight training—Ontario—History—20th century. 6. World War, 1939–1945—Canada. 7. Fighter pilots—Norway—Biography. I. Jones, Candis, 1951–, illustrator II. Title.

UG635.N6B37 2017 358.4009481 C2017-902557-0

Design and typesetting by Fox Meadow Creations
Printed in Canada by Friesens

For Fernando, Lara and Julian

Contents

Preface

In writing this book about the young Norwegian recruits who came to Canada during World War II and joined the Royal Norwegian Air Force, I discovered stories of patriotism, bravery, and sacrifice.

I was fortunate enough to personally meet with two of these former recruits, both of whom trained at Little Norway and later fought overseas with the Norwegian squadrons of Britain's Royal Air Force. I would like to thank these two veterans, Rolf Kristian Hauge of Collingwood, Ontario, Canada, and Rolf Kolling of Oslo, Norway, for generously sharing their memories with me. I would also like to thank Inger Stene, Norwegian Resistance fighter and wife of Royal Norwegian Air Force veteran, John Stene, for kindly relating her wartime experiences.

I'm grateful to family and friends of former members of the Royal Norwegian Air Force, for sharing their photos and recollections of the veterans' experiences at Little Norway. I would like to thank: Laura Anonby, Lauretta Anonby, Larry Baadsvik, Al Bacon, Barbara Chramer, John Clement, Susan Whiteside Davis, Yvonne Dixon, Eva Falkenhall, Per Falkenhall Jr., Karen Greavette, Erik Hansen, Rita Hiorth, Betty Honnor, Alan Hooper, Elaine Hooper, Sandra Howes, Randi Hvatum, J. Peter Hvidsten, Knut Larsen, Lars Lindgren, Bonnie Lund, Madeline Olsen, Pat Pedersen, Gunhild Ramm Reistad, Kjell Arne Reistad, Michael Skretteberg, and Bill Whiteside.

Many others contributed information, photos, and stories, which greatly added to this book. My thanks go to Shirley Barlow, Alice Binder, Chris Cator Everest, Knut Fossum, Cyril Fry, Marion Fry, Bo Gustafsson, Douglas Johnston, Timothy Allan Johnston, Yvonne Jones, Jason Kruusmagi, Kaye Kruusmagi, Earle Laycock, Nancy Lynch, Larry Milberry, Brian Munro, Bob Patullo, Irene Peters, Ron Stoddart, Richard Thrift, and my fellow-writers, Dona Massel and Gabriele Wills.

As always, librarians are key allies in conducting research. I'd like to thank Mette Guderud, Librarian, Armed Forces Museum Library, Oslo, Norway; Jeff Hubbell, Archivist, Ports-Toronto Archives, Toronto, Ontario; and Judy Humphries, Archivist, Gravenhurst Archives, Gravenhurst, Ontario, for their excellent contributions to my research.

Very special thanks flow to Candis Jones, Gravenhurst teacher and artist, who provided photography and photographic editing for this book—the photos bring life to the tale. Candis and I are both from Gravenhurst and so grew up a few miles from the site of Little Norway. We attended school with some of the children of Royal Norwegian Air Force members, who had returned to Muskoka after the war ended. Candis is married to Robert Olafson, son of Lieutenant Harald Olafson, who was a flight instructor at Little Norway. Candis is also the great-niece of Sergeant Svend Clement, who served with the Royal Norwegian Air Force during World War II.

Author and former mayor of Muskoka Lakes, Susan Pryke, supplied excellent editing services, and I'm grateful for her patience and insightful comments. Working with Susan was an enlightening and enjoyable experience.

Many thanks also to Marion and Cyril Fry, Gravenhurst historians, for reviewing the manuscript and for providing sage counsel throughout the writing process. I would also like to thank Lara Traficante, my daughter, for reading the manuscript and making very valuable suggestions for improvements.

I'm grateful to Gary Long, owner of Fox Meadow Creations, Sault Ste. Marie, for the outstanding design and production work done on this book and for his ongoing guidance. Gary also created the book's beautiful maps, which are so helpful in understanding of the history of the period.

And a huge thank you to my husband, Fernando Traficante, for reviewing various drafts of the manuscript and supplying excellent advice each time. Without Fernando's encouragement and the enthusiastic support of our children, Lara and Julian Traficante, this book would not have been possible.

Andrea Baston
Toronto, April 2017

Timeline

APRIL 9, 1940: Germany invades Norway and Denmark. Denmark surrenders; Norway declares war. Vidkun Quisling proclaims himself to be Norway's prime minister and the *Nasjonal Samling* party to be the country's new government.

MAY 10, 1940: Germany invades France, Holland, and Belgium.

JUNE 3, 1940: Norwegian Consul General contacts Canadian government about establishing an air training program in Canada.

JUNE 4 TO 8, 1940: Allied troops evacuate Norway. On June 7, King Haakon and Government members sail from Norway to England. The King proclaims that the struggle will be continued by the Government-in-exile.

JUNE 10, 1940: Norway surrenders to Germany.

JULY 1, 1940: Negotiations begin between Toronto Harbour Commissioners and Norwegian officials regarding use of Toronto Island Airport for flight training.

SEPTEMBER 13, 1940: Agreement signed between Toronto Harbour Commissioners and Norway regarding establishment and operation of Little Norway.

NOVEMBER 10, 1940: Official opening of Little Norway in Toronto.

FEBRUARY 17, 1941: Norwegian recruits first attend British Commonwealth Air Training Plan courses.

APRIL 16, 1941: Lieutenant Colonel Ole Reistad takes command at Little Norway.

APRIL 25, 1941: First Little Norway squadron departs for active duty overseas, becoming 330 Norwegian Squadron of the British Royal Air Force.

JUNE 20, 1941: Northrop floatplane crashes into *Sam McBride* ferry as it crosses Toronto Harbour.

JULY 1941: Elementary Flying Training School moves to Emsdale, Ontario.

JULY 21, 1941: 331 Norwegian Squadron is established in Catterick, England.

JANUARY 1, 1942: Royal Norwegian Air Force leases Muskoka Airport and one month later, purchases adjacent lands on which to build Little Norway.

JANUARY 16, 1942: 332 Norwegian Squadron is established in Catterick, England.

JANUARY 18, 1942: Official opening of Vesle Skaugum.

MAY 4, 1942: Official opening of Little Norway at Muskoka Airport. Use of Emsdale Airport for flight training terminates.

AUGUST 19, 1942: Allies launch attack at Dieppe.

APRIL 5, 1943: Royal Canadian Air Force takes over Little Norway in Toronto.

MAY 10, 1943: 333 Norwegian Squadron is established in Woodhaven, Scotland.

JUNE 6, 1944: D-Day.

NOVEMBER 10, 1944: Norwegian Army Air Service and Royal Norwegian Navy Air Service officially merge into Royal Norwegian Air Force.

FEBRUARY 17, 1945: Ceremony is held to mark closure of Little Norway at Muskoka Airport.

FEBRUARY 21, 1945: Fire at Little Norway.

MAY 7, 1945: Germany surrenders.

MAY 26, 1945: 334 Norwegian Squadron is created from 333 Squadron, B Flight, in Banff, Scotland.

JUNE 7, 1945: King Haakon VII returns to Norway.

AUGUST 14, 1945: Japan surrenders. End of World War II.

NOVEMBER 20, 1987: Official opening of Little Norway Park, Toronto.

SEPTEMBER 25, 2007: Official opening of Little Norway Memorial, Muskoka Airport.

R.N.A.F.
TRAINING CAMP
LITTLE NORWAY
: LILLE NORGE :

Introduction

> We Norwegians who are present have no more homes ... our dear ones whom we left behind us in Norway live under tyrants. But we stand here today resolved to play our part in liberating Norway.[1]

They had stolen away from their homeland to escape the Nazi invaders, survived stormy voyages across the North Atlantic in fishing boats, or travelled halfway around the world to reach Canada and resurrect Norway's air force. Now, on November 10, 1940, several hundred of these new recruits listened as Major General William Steffens spoke at the official opening of camp "Little Norway" in Toronto.

The young men wanted to train as pilots, gunners, navigators, and aircraft mechanics, then return to Norway to defeat the German military forces that had invaded their country.

Canada would give the young Norwegians that opportunity.

The story

Exile Air begins with Germany's invasion of Norway, then briefly describes the subsequent 60-day war, where Norwegian and Allied military forces fought the German troops occupying the country.[2] Norway was forced to surrender on June 10, 1940, but many of its young patriots escaped from the country over the coming years and joined the Royal Norwegian Air Force training at Little Norway in Toronto and, subsequently, in Muskoka. This is their story.

OPPOSITE Little Norway, Royal Norwegian Air Force training facility, November 1940.
PHOTO: ARCHIVES OF ONTARIO, C 109-18-11-35 H

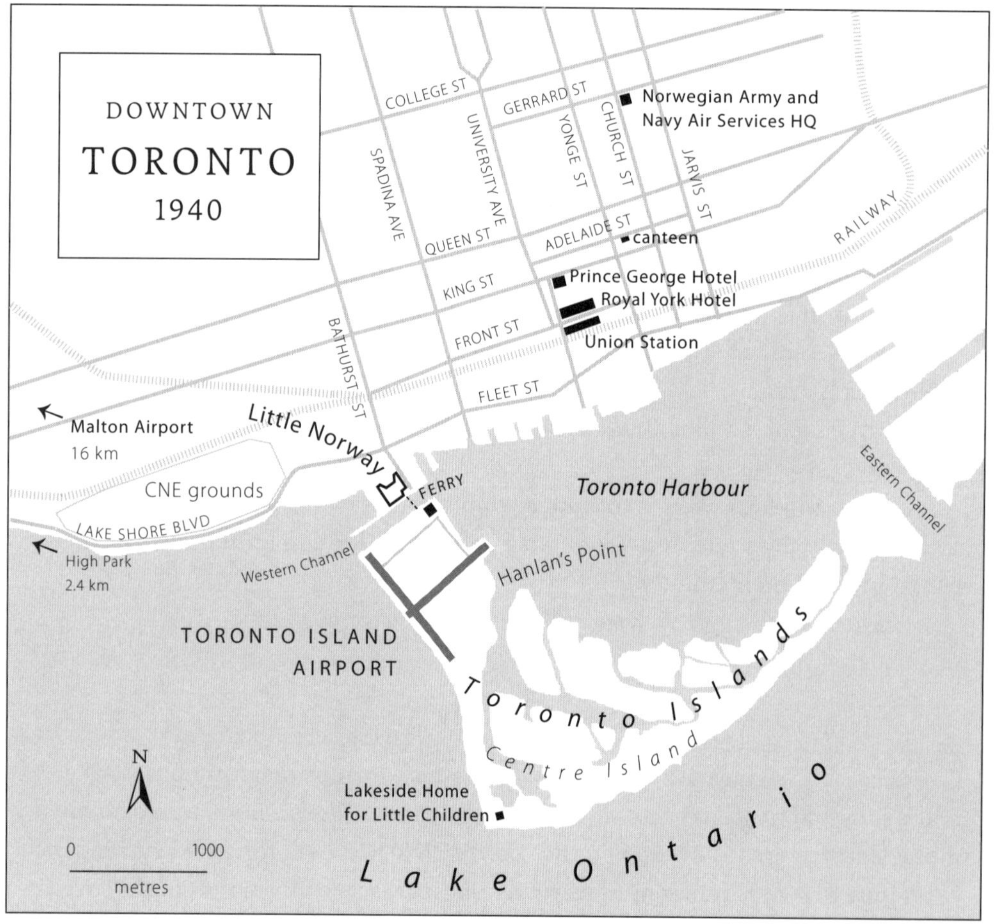
DOWNTOWN
TORONTO
1940
COLLEGE ST
GERRARD ST
Norwegian Army and
Navy Air Services HQ
SPADINA AVE
UNIVERSITY AVE
YONGE ST
CHURCH ST
JARVIS ST
QUEEN ST
ADELAIDE ST
canteen
RAILWAY
KING ST
Prince George Hotel
Royal York Hotel
FRONT ST
Union Station
BATHURST ST
FLEET ST
Malton Airport
16 km
Little Norway
CNE grounds
LAKE SHORE BLVD
FERRY
Toronto Harbour
Eastern Channel
High Park
2.4 km
Western Channel
Hanlan's Point
TORONTO ISLAND
AIRPORT
Toronto Islands
Centre Island
N
Lakeside Home
for Little Children
0
1000
metres
Lake Ontario

MUSKOKA
1941–1945
N
0 10
kilometres
EMSDALE AIRFIELD
Emsdale
ALGONQUIN
PARK
Long (Dotty) Lake
Vesle Skaugum
Oxbow Lake
road
CN RAILWAY
Huntsville
Limberlost Lodge
Fairy Lake
Lake of Bays
Port Carling
CN RAILWAY
Bracebridge
Lake
Muskoka
Bala
MUSKOKA AIRPORT
Little Norway
Muskoka Beach
Reay
Gravenhurst
to Toronto
Emsdale
Vesle Skaugum
Georgian
Bay
Muskoka
Airport
Orillia
L Simcoe
Barrie
0 50
km
Toronto
Lake
Ontario

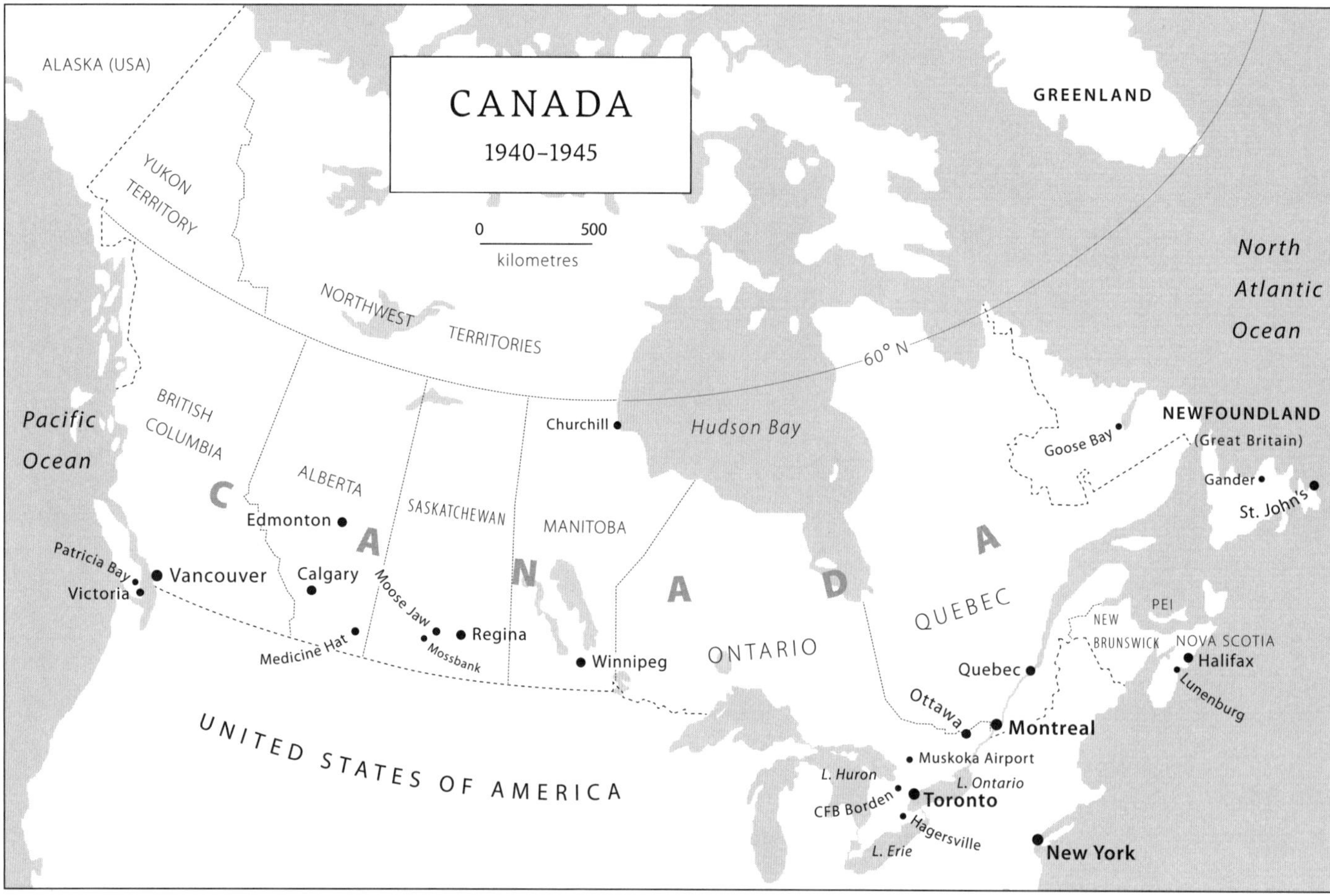
CANADA
1940–1945
0
500
kilometres
ALASKA (USA)
YUKON
TERRITORY
NORTHWEST
TERRITORIES
GREENLAND
North
Atlantic
Ocean
60° N
Pacific
Ocean
BRITISH
COLUMBIA
ALBERTA
SASKATCHEWAN
MANITOBA
Churchill
Hudson Bay
Goose Bay
NEWFOUNDLAND
(Great Britain)
Gander
St. John's
C
A
N
A
D
A
Edmonton
Calgary
Patricia Bay
Vancouver
Victoria
Medicine Hat
Moose Jaw
Mossbank
Regina
Winnipeg
ONTARIO
QUEBEC
NEW
BRUNSWICK
PEI
NOVA SCOTIA
Halifax
Lunenburg
Quebec
Ottawa
Montreal
Muskoka Airport
L. Huron
L. Ontario
CFB Borden
Toronto
Hagersville
L. Erie
New York
UNITED STATES OF AMERICA

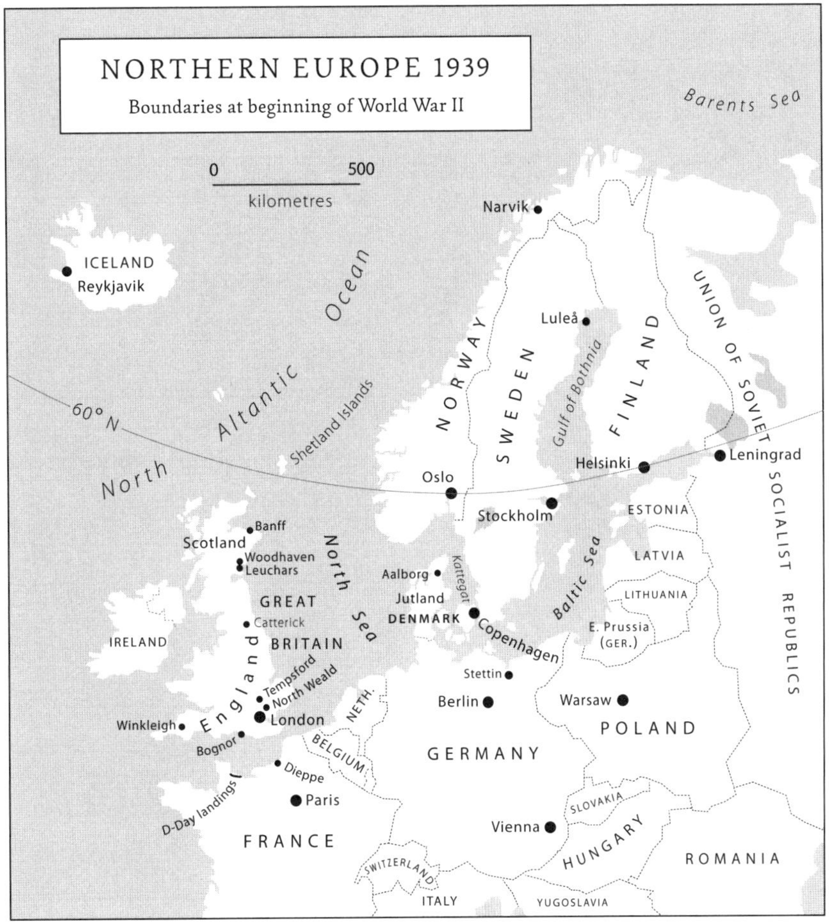
NORTHERN EUROPE 1939
Boundaries at beginning of World War II
0
500
kilometres
Barents Sea
Narvik
ICELAND
Reykjavik
North Atlantic Ocean
60° N
Shetland Islands
NORWAY
SWEDEN
FINLAND
UNION OF SOVIET SOCIALIST REPUBLICS
Luleå
Gulf of Bothnia
Helsinki
Leningrad
Oslo
Stockholm
ESTONIA
LATVIA
LITHUANIA
E. Prussia
(GER.)
Baltic Sea
Kattegat
Aalborg
Jutland
DENMARK
Copenhagen
Banff
Scotland
Woodhaven
Leuchars
North Sea
GREAT
BRITAIN
Catterick
IRELAND
England
Tempsford
North Weald
London
Winkleigh
Bognor
NETH.
BELGIUM
Dieppe
Paris
D-Day landings
FRANCE
Stettin
Berlin
GERMANY
Warsaw
POLAND
Vienna
SLOVAKIA
HUNGARY
ROMANIA
SWITZERLAND
ITALY
YUGOSLAVIA

NORWAY
1940
German invasion targets
(*Weserübung*)
0
100
200
kilometres
Kirkenes
Finnmark
Tromsø
Bardufoss
Petsamo
U.S.S.R.
Elvegårdsmoen
Narvik
Lofoten Islands
FINLAND
SWEDEN
NORWAY
Luleå
Ocean
Atlantic
North
65° N
Namsos
Nordli
Trondheim
Kristiansund
Tustna
Gjemnes
Molde
Støren
Åndalsnes
Romsdalshorn
Lake Lesjaskog
Røros
Lake Aursunden
Gudbrandsdalen
Glomma River
Østerdalen
Gulf of Bothnia
Nybergsund
Lillehammer
Lundebygda
Hamar
Elverum
Bergen
Televaag
Lake Tinnsjø
Bjørgeseter
Hakadal
Kongsvinger
Vemork
OSLO
Magnor
Kongsberg
Horten
Sarpsborg
Stavanger
Sola
STOCKHOLM
Baltic Sea
Egersund
Kristiansand
Arendal
Mandal
Hakadal
Lake Steinsfjorden
Lysaker Bridge
Kjeller
Lommedalen
Fornebu
OSLO
Drammen
Oslofjord
Oscarsborg Fortress
N
0
20
km
Horten
Sarpsborg

1

An Uncomfortable Neutrality

At first, Hitler hadn't wanted to invade Norway. He had to be persuaded.

After German forces attacked Poland in 1939 and Britain and France declared war on Germany in response, Norway proclaimed itself to be a neutral country—Norway hadn't fought in World War I or any other war for 125 years.

Hitler accepted Norway's neutrality, along with that of Sweden and Denmark. A non-aligned Scandinavia didn't appear to threaten his country. Hitler also recognized the challenges arising from any military action against Norway: an attack from the sea would be highly dangerous, undoubtedly leading to battles between the German Navy and the superior British Royal Navy. Even if the Germans won the conflict and occupied Norway, its mountainous interior, frigid winters, and snow accumulation would place a significant strain on German resources.

But Norway's neutrality didn't sit well with German naval leaders or, for that matter, with a number of British military advisors, for Norway was involved in the supply of a crucial ingredient to Germany's war industries: iron ore.[1]

Iron ore is used in the manufacture of steel, a vital component for production of military tanks, planes, and ships. Germany imported most of its iron ore from Sweden. During the warmer months, the iron ore was shipped from Luleå in northern Sweden and travelled across the Baltic Sea to Germany. However, during the colder months, much of the Baltic Sea was frozen so Sweden transported the iron ore by rail to Narvik, Norway. From Narvik, German merchant ships carried the iron ore south through Norwegian territorial waters to Germany.

Hitler's naval advisors had heard reports for several months that British troops would soon land in Norway to stop the iron ore transport. If Britain occupied Norway, not only would Germany lose the iron ore supplies travelling by sea from Narvik during the winter, but Sweden might be drawn into the war on the Allies' side and cease all iron ore exports to Germany. Grand Admiral Erich Raeder, Commander in Chief of the German Navy, warned Hitler to seize Norway before Britain did.[2]

Vidkun Quisling

Vidkun Quisling, a man whose name would later become synonymous with the word "traitor," shared Hitler's advisors' opinion. Quisling was a prominent Norwegian fascist. He had been a member of the Norwegian Labour Government as its minister of defence from 1931 to 1933. A Norwegian general, who knew Quisling then, described him as being out for himself. The general said Quisling was "ambitious and unstable—dangerous qualities when they are found together."[3]

Portrait of Vidkun Quisling.

PHOTO: NATIONAL ARCHIVES OF NORWAY.

When the Labour Government fell, Quisling established the *Nasjonal Samling* or National Unity Party, a political party which copied the Nazis' style, methods, and ideology.

The Nasjonal Samling did poorly in the 1936 Norwegian parliamentary elections, obtaining less than two per cent of the popular vote. Quisling lost his own attempt to win a seat in the *Storting*, Norway's Parliament.

He travelled to Germany in December 1939 to gain support for the Nasjonal Samling, meeting with Hitler a couple of times and promoting his belief that Britain intended to land troops in Norway imminently. Quisling wanted to stage a coup: his supporters would overthrow the existing Norwegian government and replace it with a Nasjonal Samling government. Then as prime minister, Quisling would formally invite the Nazis to occupy Norway.

Hitler listened to his naval advisers and Quisling. At one of his meetings with Quisling, he repeatedly said that he personally would prefer to keep Norway and all of Scandinavia completely neutral, but he could never permit the enemy to occupy those countries.[4] In the end, Hitler ordered his staff to consult with Quisling and work out a plan for the German invasion of Norway.

British plans

Hitler's naval advisers were right. Britain did want to stop the flow of Swedish iron ore to Germany.

Winston Churchill, First Lord of the Admiralty, believed that depriving Germany of a continued supply of iron ore would severely damage Hitler's war effort. Early in the war, Churchill had begun lobbying the British War Cabinet to lay mines in Norwegian territorial waters, so German iron ore transports would have to sail farther out to sea where they could be attacked by British warships. However, Prime Minister Neville Chamberlain and the War Cabinet didn't approve the mine-laying plan.

The situation changed when the Soviet Union made a claim to some Finnish territory and invaded Finland on November 30, 1939. Britain and France wanted to help the Finns battle Soviet aggression and began preparing to send a military force there. To reach Finland, Allied troops would first have to land in Norway and then travel through Sweden to the war zone.

The Allies realized that, as well as supporting the Finns, this initiative would provide a unique opportunity to halt the iron ore traffic to Germany. Secret plans were developed, whereby most of the force would land in Narvik and travel by train across Scandinavia to fight in Finland. However, some troops would remain in Narvik, occupy that city, and stop shipments of iron ore from leaving for Germany during the winter months. Other Allied troops would seize the Swedish port of Luleå, and prevent iron ore deliveries to the enemy during the summer months.

The Allies didn't expect Hitler to calmly accept their landings in Norway, purportedly to assist Finland; they thought he would likely react by launching an attack on Scandinavia. If he did, the German offensive would provide the Allies with another occasion to obtain a foothold in Norway. Should Germany strike at Norway, British and French troops would occupy Norway's western ports, and block the iron ore traffic with the Nazis. Prime Minister Neville Chamberlain later told the British House of Commons that any occupation of Norway by Allied forces was only to occur in one of two situations: either the troops were invited to occupy the ports by Norway, or Germany had already violated Norway's neutrality.[5] In preparation for action, the Allies readied a small invasion force.

But all the plotting and planning came to nothing. Britain made preliminary requests to Norway and Sweden for consent to land Allied troops in their territories and to convey the troops to Finland. Both Norway and Sweden absolutely refused permission. Still, the Allies continued quietly planning for the operation until March 12, 1940, when Finland surrendered to the Soviets.

The reason for the Allies' proposed landings in Norway disappeared with the end of the Finnish-Soviet war and as a result, British and French government leaders began to pay more attention to Churchill's mine-laying proposals.

Trigger

One event triggered Hitler's decision to invade Norway. On February 16, 1940, the German supply ship *Altmark* was sailing through Norwegian territorial waters on its way home. The ship carried a secret cargo: 299 seamen, mostly British, captured from merchant ships sunk by the German heavy cruiser *Graf Spee*. These prisoners were kept in the *Altmark*'s dark, watery hold and given little food. Reportedly, one desperate prisoner found a piece of glass and pried open a porthole. At night, he would hold the glass outside the porthole and try to catch the moon's reflection, hoping the British would see the light and rescue them.[6]

Norwegian naval authorities visited the *Altmark* on three occasions while it remained in Norway's waters, each time performing a cursory inspection. The *Altmark*'s captain assured the Norwegian authorities that the vessel was just a merchant ship carrying oil and should be allowed to proceed on its course. According to one of the prisoners, the men in the hold tried to attract the Norwegians' attention by banging on the ship's sides with stanchions. The crew of a Norwegian destroyer sailing close to the *Altmark* reported hearing SOS whistle signals and seeing white handkerchiefs waving at portholes.[7] To drown out the sounds, the Germans started up winches and other machinery.

Regardless, the Norwegian authorities gave their approval for the ship to continue its journey to Germany.

British Royal Air Force (RAF) planes flying over the area

Weserübung, German destroyers at the quayside, April 6, 1940.

PHOTO: FEDERAL ARCHIVES OF GERMANY, BILD 101II-MW-5607-03. PHOTOGRAPHER: LANDER

spotted the *Altmark* and alerted the Royal Navy, which dispatched a flotilla of six destroyers to Norway. Winston Churchill became personally involved in the matter, ordering that the *Altmark* be boarded and its prisoners released.

As a result, the British destroyers intercepted the *Altmark* and troops boarded it. Shots echoed through the ship. In the skirmish, some four to seven German crew members were killed, the numbers varying with the reports of the incident.[8] One prisoner told the *Toronto Daily Star* that he would never forget how he felt when the hatch opened and someone yelled, "Any British down there?"[9]

Norway reacted to the incident by protesting heatedly at Britain's violation of Norwegian territorial waters. Britain replied that Norway had breached its own neutrality by allowing Germany to use its waters for the transfer of Allied prisoners of war to a German prison.

An enraged Hitler concluded that Britain wouldn't respect Norway's neutrality and that Norwegian territorial waters were no longer safe for iron ore shipping to Germany. Also, both Germany and Britain realized that Norway would be unlikely to react militarily to contraventions of its sovereignty.

"Weserübung"

On March 1, 1940, Hitler directed the invasion of Norway in an initiative called Operation *Weserübung*, or Weser-Exercise, named after Germany's Weser River. In the directive, he stated that Weserübung's goal was to prevent British encroachment in Scandinavia and the Baltic, and to guarantee Germany's ore base in Sweden. Further, it would give the German Navy and Luftwaffe a wider start-line against Britain.[10]

Hitler ordered that the invasion be portrayed as a peaceful occupation, intended to protect Scandinavia's neutrality from the British; but, despite this, Hitler wanted any resistance to be crushed by all military means. The Royal Navy posed a threat to the German fleet should the invasion preparations be discovered, so Hitler stressed that the attack had to take Norway and the Allies by surprise.

Denmark, which was viewed as a connecting bridge to Norway, was to be invaded simultaneously.[11] As a bonus, Germany would gain excellent Danish airfields, such as Aalborg in Jutland. From these airfields, enemy airplanes could attack Britain.

General Nikolaus von Falkenhorst, then commander of the German Army's 21st Corps, was appointed leader. Hitler named April 9 as the date for the Weserübung.

The plan was for seven cruisers, 14 destroyers, numerous torpedo boats and small vessels to be divided into six *Gruppen* (groups), bringing an advance assault force of about 10,000 troops to Norway. Each group of vessels would target a major Norwegian port. Gruppe 1 would target Narvik; Gruppe 2, Trondheim; Gruppe 3, Bergen; Gruppe 4, Kristiansand and Arendal; Gruppe 5, Oslo; Gruppe 6, Egersund. Group 5 was ordered to capture the King of Norway, Haakon VII, along with members of government and important military and civic leaders, before local resistance could be organized.

The Luftwaffe's 10th Air Corps, flying 1,000 planes of various types, would take part in the invasion, protecting the German Navy and transporting troops to seize airfields at Oslo, Kristiansand, and Stavanger.[12]

All naval forces were to arrive at the entrance to their target cities' harbours at about 4:15 a.m. Norwegian time and the attacks were to be carried out within a half hour of arrival.

The first German warships set sail on April 3, heading for Norway's most northerly ports.

The Royal Navy during the Second World War. The wire stays are removed from the mines on board the minelayer HMS *Apollo*, in enemy waters off Norway. PHOTO: IMPERIAL WAR MUSEUM, WWII IWM A 27055.

Mines in Norwegian waters

On April 5, Churchill finally received approval from the Allied Supreme War Council to lay mines in Norwegian waters. The same day, the Allies notified Norway and Sweden that they reserved the right to take such measures as they might think necessary to prevent Germany from obtaining resources from the two neutral countries that would give it an advantage in the war.

The British mines were scheduled to be laid on April 8. Anticipating that Germany would react to the mine-laying by invading Norway, a small number of Allied troops boarded

Norway's Defences

For years, the Norwegian government's spending on defence had been low, partly due to a pacifist attitude common across Europe after World War I; fortresses protecting fjords were neglected, and military equipment was often faulty, incomplete, or out of date.

Norway didn't possess a separate air force. Instead, the army and navy each had an airborne division, the Norwegian Army Air Service (Army Air Service) and the Royal Norwegian Navy Air Service (Navy Air Service). Most of the aircraft flown by both services were antiquated. As of 1938, the Army Air Service possessed 4 Caproni Ca 310 bombers/reconnaissance aircraft, some 45 Fokker C.V reconnaissance biplanes, and 38 training planes. The Navy Air Service had 8 Douglas DT torpedo bombers, 17 Mariens Flyvebaatfabrik M.F. 11 reconnaissance biplanes, and 11 training planes.[14]

However, as the threat of war spread throughout Europe, Norway had purchased some relatively modern planes to increase its military strength, including 12 Gloster Gladiator biplane fighters from Britain and 6 Heinkel He 115, twin-engine torpedo bomber floatplanes from Germany. In the months preceding Germany's attack, Norway had placed orders with American and Italian aircraft manufacturers for newer planes. Nineteen of 24 American Curtiss P-36 Hawk fighters had been delivered by April 8, but none was operational.

British ships on April 7; if Germany attacked Norway, these troops were to land and hold western Norwegian ports until reinforcements arrived.

And so, in a curious coincidence, soldiers from both Germany and Britain were sailing towards Norway at about the same time—though the British troops in far fewer numbers.

Early in the morning of April 8, British and French warships mined Norwegian territorial waters off Narvik. The Allies notified Norway afterwards, advising that the mines were intended to obstruct ships carrying war contraband, such as weapons or ammunition, and to force these vessels out to the high seas where they could be intercepted. Britain said it was prepared for any action Germany might take and would react sharply to any German military action against Scandinavia.[13]

The Norwegian Cabinet debated their response to Britain's mine-laying all morning and the final decision was to strenuously protest the Allies' actions and demand that the mines be removed.

In the meantime, a German invasion was steaming its way towards Norway.

2

Invasion

While Norway's Government was dealing with the mine-laying issue on April 8, a number of alarming events were taking place.

Stories had spread in Europe at the end of March 1940, that German troops and ships were massing in coastal cities and soldiers' leaves had been cancelled. For the most part, however, these early warnings were ignored.

On April 7, the Norwegian Embassy in Berlin reported that 15 to 20 troop transports had left Stettin, a city on the north coast of Germany, and were proceeding west to an unknown destination.

Early on April 8, a Polish submarine sank a German transport ship, the *Rio de Janeiro*, in international waters near Kristiansand, Norway. Some 200 troops wearing German uniforms were aboard, along with horses and military equipment. One soldier told his rescuers the ship was en route to Bergen to help defend it against the British.[1]

That same morning, the British destroyer *Glowworm*, which had provided protection for the Allies' mine-laying activities near Narvik a few hours earlier, came across two German destroyers and fired on them. One of these was the heavy cruiser *Admiral Hipper*, on its way to invade Trondheim. The *Admiral Hipper* sank the *Glowworm*.

In the afternoon, a British submarine destroyed a German tanker, the *Posidonia*, sailing in international waters near the entrance to Oslofjord, the 120-kilometre stretch of water leading to Oslo.

Foreign planes were sighted flying out at sea in several locations along Norway's coast and circling over Norwegian cities. Reports circulated of 15 to 90 German warships heading north through the Kattegat sea towards Norway.[2]

But even in the face of all these threats, Norwegian and British military leaders did little. A Norwegian admiral ordered some coastal lighthouses to be extinguished and called up extra staff to guard certain fortresses, but no major actions were taken, such as mobilizing the Norwegian Army.

Many Allied military leaders didn't believe that Hitler would ever attack Scandinavia and risk battling the Royal Navy. Some thought Hitler was just organizing his forces for a counterstrike should the Allies occupy Norwegian ports. Members of Norway's naval staff concluded that the Germans intended to attack the Netherlands.

Around 6:00 p.m., the British Admiralty sent a telegram to the Norwegian Embassy advising that German ships had been observed in the North Sea: "It is assumed for certain that the goal is to undertake operations against Narvik and they could arrive there before midnight."[3]

Still, Norway's military officials took no major defensive action.

Germany's ultimatum

Germany's Minister for Norway, Dr. Curt Bräuer, called Norway's Foreign Minister, Halvdan Koht, around 4:00 a.m. on April 9; at a subsequent meeting, Bräuer presented Koht with a memorandum. A similar memorandum was given to Denmark's Foreign Minister in Copenhagen at about the same hour.

The document alleged that Britain and France were guilty of breaches of international law, and planned to invade and occupy Norway, and that the Norwegian government intended to acquiesce to the Allied occupation. The memorandum also stated that German troops had not come as enemies, but to protect the country from the pending Anglo-French occupation. Germany expected that the Norwegian government and people would offer no resistance. Any resistance would be broken by all possible means and would lead only to "absolutely useless bloodshed."[4]

Meanwhile, Norway and Denmark were under attack.

Oslo

Hitler had charged Gruppe 5, the German flotilla approaching Oslofjord, with taking the city and capturing King Haakon VII, members of Norway's Government, and others.

The brand-new, 18,000-ton, heavy cruiser *Blücher* led the group of ships, accompanied by a second heavy cruiser, *Lützow*, a light cruiser, *Emden*, a torpedo boat, *Albatros*, and other craft.

Given the long distance from the fjord's entrance to Oslo, it was difficult for such a large flotilla to reach its goal without detection. Around 11:00 p.m. on April 8, the *Pol III*, a Norwegian patrol boat, sighted the German flotilla. The *Albatros* ordered the *Pol III* to surrender, but its captain, Leif Welding-Olsen, refused and fired flares to alert Norwegian gun batteries a little further along the fjord to the presence of the intruders. In the ensuing skirmish, the *Pol III* rammed the *Albatros* and the German torpedo boat fired back, wounding Welding-Olsen in both legs. The crew abandoned the burning ship and were captured, but Welding-Olsen, weakened by blood loss, drowned in the fjord. He became the first Norwegian to die in the war between Germany and Norway.

Gruppe 5 continued up the fjord. A small German landing force boarded escort boats and headed for the town of Horten and its nearby naval base. A Norwegian minelayer, the *Olav Tryggvason*, awaited the Germans in Horten harbour and the minelayer sank one of the escort boats; however, that didn't prevent another boat from landing troops, who soon captured both the naval base and Horten.

The *Blücher* was drawing near to Oscarsborg, an aging fortress located on two islands in Oslofjord. Oscarsborg's commander, Colonel Birger Eriksen, had requested funding to purchase modern weapons for the fortress many times, but the government hadn't supplied much. Oscarsborg's battery of guns was over 40 years old; no one knew whether the fortress's torpedoes would operate properly as they hadn't been fired for a very long time.

Eriksen had been warned that an unidentified flotilla was proceeding up Oslofjord. As Norway was a neutral country, Eriksen wasn't sure if he should fire the guns: was a warning shot first required under the rules of engagement, he wondered? But the situation required a quick decision. As Eriksen had been unable to obtain orders from his superior officers, he decided to fire the guns on the intruding ships, saying that he would either be decorated or court-martialled.[5]

As the *Blücher* reached Oscarsborg, the fortress's main gun batteries fired, seriously damaging the heavy cruiser. Still, the boat continued until it neared Oscarsborg's underwater torpedo battery, located on the small island behind the main fortress. Two torpedoes from the battery struck the *Blücher*.

According to a German soldier who was later captured, the troops on the *Blücher* hadn't been told at first where they were heading. When on the high sea, German officers advised the troops they were sailing for Norway in response to that country's request for protection from the British threat. The soldier recalled the heavy detonations shaking the ship:

> The bulkhead of the deck on which we were standing was smashed in. I heard cries from comrades who had been wounded. The fumes of burning oil were surrounding us. The engines had been hit. The infantrymen gathered on deck could hardly keep their balance. Medical squads were carrying away the wounded.... Finally came the last command, "Everybody jump."... The water was icy. The 300 yards that separated us from the islands seemed a terrible distance to me. The cold numbed me. The clothes I hadn't had time to get rid of hindered my movements, but I had to swim without stopping to think, or else it meant death.[6]

The German soldier scrambled onto an island and looked back to see the *Blücher* ablaze. With black smoke billowing, it sank into Oslofjord.

There are inconsistent views as to the number of lives lost, ranging from around 1,000 to 1,600.[7]

Oscarsborg's artillery battered the *Lützow* and it, along with the remaining German warships, withdrew further off from the fortress. Once the ships were sufficiently distant, they made for the east side of Oslofjord and landed troops, who began to march towards Oslo.

The German cruiser *Blücher* listing heavily to port after being hit by cannon fire and torpedoes from the Norwegian coastal fortress Oscarsborg. The *Blücher* sank a short time later. PHOTO: NATIONAL ARCHIVES OF NORWAY.

Later that day, waves of Luftwaffe aircraft bombed the old fortress from above. In the early morning hours of April 10, Oscarsborg surrendered.

But the actions of Colonel Birger Eriksen and the guardsmen at Oscarsborg had an enormous impact on Hitler's plans. By sinking the *Blücher*, the capture of Oslo was delayed by a sufficient amount of time for King Haakon VII, Government members, and protectors of the country's gold supply to escape from the city.

German soldiers marching through Oslo on April 9, 1940.
PHOTO: *VARDEN*.

Fornebu Air Base

The Army Air Service's only fighter squadron flew from Fornebu Air Base, some 10 kilometres from Oslo. When Germany attacked, the fighter squadron had 10 Gloster Gladiator biplanes, but only five of them were operational.

The pilots had spent an unsettled night, listening to sirens sounding from Oslofjord. No one knew exactly what was occurring: one pilot later wrote that civilians, listening to regular radio broadcasts, had better information that morning than the Norwegian Air Services did.[8]

The men were roused from their beds before dawn and taken to Fornebu, where their commanding officer told them that unknown planes were coming from the south, flying north. Two Gladiators were sent off on patrol, one flown by Second Lieutenant Finn Thorsager, an Oslo native. Thorsager spotted a German Dornier aircraft and fired at it numerous times before it disappeared into the clouds.

A second patrol by other pilots at 6:00 a.m. returned to base without glimpsing anything. However, their luck soon changed. About 7:00 a.m., Lieutenant Rolf Torbjørn Tradin led the five Gladiators, including one flown by Thorsager and one by Sergeant Per Waaler, back up into the sky. Thorsager noticed a couple of other Gladiators being worked on by mechanics on the ground, in the hopes of making them operational.

Flying over Oslofjord, the pilots could see a plume of thick, black smoke ascending from the sunken *Blücher*. Before long, the pilots spotted German aircraft flying below them: wave after wave of enemy planes, too numerous to count. Realizing the handful of Gladiators was drastically outmatched and outnumbered by the Germans, Tradin radioed to the Norwegian pilots: "Break formation. Every man for himself. Good luck!"[9]

In the ensuing dogfights, Waaler is believed to have shot down a Heinkel He 111 bomber, the first enemy plane destroyed over Norwegian soil. Thorsager, or Thorsager jointly with Tradin, probably destroyed another Heinkel.[10]

The air battle continued for some time and one Norwegian Gladiator was shot down. The Luftwaffe destroyed two other

Gladiators, which were on the ground at Fornebu. The four remaining Gladiators escaped to the countryside, but none was ever fit for combat again.

Estimates of the number of German aircraft destroyed in the dogfights at Fornebu that morning range from four to six.[11]

Norwegian machine-gunners on the airfield destroyed a German Junkers JU 52 troop transport. Regardless, other transports soon delivered German paratroopers and soldiers who, after occupying the airfield, began to march towards Oslo.

The enemy troops reached Oslo and by 2:00 p.m., the city capitulated.

Kjeller Air Base

Captain Ole Imerslun Reistad commanded the *Speidervingen* (Reconnaissance Wing) of the Army Air Service at Kjeller Air Base, some 22 kilometres northeast of Oslo. At 41 years of age, Reistad had been an Olympic athlete and was very well known in Norway.

On April 9, nine Fokker C.V biplanes were parked on the airfield at Kjeller Air Base.

Hearing rumours on April 8 of German ships sailing off the coast of Norway and foreign aircraft flying over Norwegian territory, Reistad realized that an attack of some kind could be imminent and that Kjeller's outdated Fokkers would be no opposition for the faster enemy fighters. He tried to reach his commanding officers for instructions, but received no clear orders.

Very early in the morning of April 9, Reistad decided that the wing's aircraft should depart from Kjeller as rapidly as possible. He ordered that the biplanes' wheels be replaced with skis, so they could land on frozen lakes in the countryside. Luckily, many of the Fokkers were ready to go.

Front view of a Fokker biplane.

PHOTO: STATE ARCHIVES OF NORTH CAROLINA.

Seven aircraft, including one flown by Reistad himself, took to the skies. The Fokkers landed on Lake Steinsfjorden, northwest of Oslo, where they were soon joined by other escaping aircraft. This group would move from lake to lake over the following days to avoid German aerial attacks and to support the ground fighting by Norwegian troops.[12]

German troops disarm Norwegian soldiers.

PHOTO: FEDERAL ARCHIVES OF GERMANY, RH 82 BILD-00095. PHOTOGRAPHER: MAX EHLERT.

Narvik

Weserübung's Gruppe 1, composed of 10 German destroyers divided into three flotillas, battled heavy seas as it raced up Ofotfjord, the 78-kilometre waterway leading to Narvik.

One flotilla landed a battalion of German soldiers northeast of Narvik to capture an ammunitions and supply depot. Heavily outnumbered, the 17 Norwegian Army soldiers guarding the depot surrendered to the invaders.

Norway had two coastal defence ships, the *Eidsvold* and the *Norge*, both antiquated, and both moored in Narvik's waters under the overall command of Captain Per Askim. The *Eidsvold* was anchored outside the entrance to the harbour while the *Norge* was anchored within its waters. A Norwegian patrol boat alerted them to the presence of the invading ships in the fjord.

As three German destroyers entered the harbour, the *Eidsvold*'s captain, Odd Isaachsen Willoch, fired a warning shot and raised signal flags ordering the foreign ships to identify themselves.

Commodore Friedrich Bonte, commander of Gruppe 1 and sailing in the flagship *Wilhelm Heidkamp*, signalled back that he was sending over a boat with an officer. In the meantime, the other two German destroyers continued sailing towards Narvik.

An officer dispatched from the *Wilhelm Heidkamp* boarded the *Eidsvold* and advised Willoch that the Germans had come as friends to defend Norway from Britain. He then demanded that the *Eidsvold* be surrendered.

Willoch asked for 10 minutes to contact his superior, Captain Askim, for instructions. Although the German officer refused, the *Eidsvold*'s radio operator had already reached the

Norge and received orders from Askim to open fire. Willoch told the German officer that the *Eidsvold* would resist.

The German officer saluted, left the *Eidsvold* and, while returning to the *Wilhelm Heidkamp*, shot a flare into the sky. The flare signalled that the Norwegians had not surrendered their ship.

The *Eidsvold* headed towards the *Wilhelm Heidkamp*. When the ships were only 300 metres apart, the *Wilhelm Heidkamp* fired its torpedoes and three of these hit the *Eidsvold*. The ship broke in two and sank within 15 seconds. Captain Willoch and 175 of the 181 crew members drowned.[13]

Meanwhile, the two German destroyers had made their way into Narvik's harbour. When the destroyers suddenly materialized from the dark, the *Norge* fired its guns. The German ships launched torpedoes and two struck the *Norge*; it sank within a minute of impact. Captain Askim was dragged unconscious from the sea and survived, but 101 of the 191 crew members were lost.[14]

With the sinking of the two Norwegian coastal defenders, the German destroyers proceeded to land troops, who made a rapid advance into Narvik. The commander of the local Norwegian Army regiment surrendered the city without a fight. It had taken the Germans two hours to capture Narvik.

Kristiansand, Arendal, Egersund

Gruppe 4, targeting Kristiansand and led by the light cruiser *Karlsruhe*, met fierce resistance from the fortress on Odderøya Island. Twice the German warships had to withdraw from the siege to avoid destruction.

However, following Luftwaffe bombing attacks on the fortress and the landing of paratroopers at nearby Kjevik Airfield, Odderøya's resistance ended and Kristiansand was soon occupied.[15]

The swastika is hoisted amid great ceremony at Akershus, April 9, 1940.
PHOTO: NATIONAL ARCHIVES OF NORWAY.

The coastal towns of Arendal and Egersund served as terminals for underwater telegraph cables running to Britain; Hitler wished to have these communication sources shut down. So, while most of the German ships in Gruppe 4 attacked Kristiansand, one torpedo boat made its way to Arendal and landed troops there.

Gruppe 6, the smallest in Weserübung, sailed to Egersund. Soon German troops controlled all three communities.

Bergen

Kvarven Fortress at Bergen fired on the Gruppe 3 flotilla, seriously damaging the light cruiser *Königsberg*, and the naval artillery training ship *Bremse*. However, soldiers disembarking from other ships captured the city before noon.[16]

Trondheim

The city of Trondheim was taken by troops from Gruppe 2's heavy cruiser, the *Admiral Hipper*, and four destroyers. Although the city surrendered without resistance, some forts fought for a few hours.

The Reconnaissance Squadron in the Trøndelag Air Unit was based at Værnes Air Station near Trondheim. Airmen there, including 22-year-old Deputy Commander Wilhelm Mohr, equipped some nine Fokker aircraft with skis and flew the planes up and away from the Air Station before the Luftwaffe arrived.[17]

Sola Air Station

German paratroopers seized Norway's biggest air station, Sola, located close to Stavanger. This action gave Germany immediate air dominance, as bombers taking flight from Sola could reach England's northern cities and attack Allied vessels sailing in the North Sea.[18] A handful of Fokker and Caproni aircraft managed to escape from Sola in advance of the attack.

Enemy troops met no resistance in taking the city of Stavanger.

Denmark

On April 9, 1940, at 4:15 a.m. Norwegian time, Hitler's ground troops crossed over the border between Germany and Denmark in motorized vehicles and quickly moved up the Jutland Peninsula. Minor skirmishes with Royal Danish Army defenders broke out in scattered locations in Jutland, but were speedily determined in the Germans' favour.

Around the same time as the ground assault was taking place, the German Navy was landing troops at a number of Danish ports, including Nyborg on Funen Island and Kørsor on Zealand Island. The soldiers who alighted at Kørsor began marching across the island towards Copenhagen, Denmark's capital city.

Hundreds of paratroops arrived at Aalborg Airfield in northern Jutland, where they met no resistance. Before long, German troop transports and other aircraft were safely landing there.

Meanwhile, the troopship *Hansestadt Danzig* and escort ships sailed into Copenhagen harbour, where soldiers disembarked. About 800 troops captured the citadel, Kastellet, an ancient, star-shaped fortress in Copenhagen, without much trouble.

However, German troops arriving at Amalienborg, home to Denmark's Royal Family, received a different reception from the King's Royal Guard. The first German attack was beaten back, allowing King Christian X, the Government, and military leaders time to decide about next steps. Meanwhile, Luftwaffe bombers roared low over Copenhagen, dropping leaflets advising the population to keep calm and co-operate with the Germans. The leaflets said that Germany had invaded

Denmark to counteract a British-French plan to expand the war to Scandinavia.

Faced with Hitler's threat of bombing Copenhagen if Denmark didn't surrender, King Christian X and the Government capitulated after a few hours and ordered the population to cease all resistance. Denmark agreed to accept Germany's protection, under protest, and was promised territorial integrity and political independence. German Major General Kurt Himer, Chief of Staff, met a few hours later with the 70-year-old King; he described the King as appearing inwardly shattered, his whole body trembling, but outwardly, absolutely dignified. He asked to retain his bodyguard and Himer replied that no doubt the Fuehrer would agree to that. The King was visibly relieved and, at the end of the interview, he said to Himer: "General, may I, as an old soldier, tell you something? As soldier to soldier? You Germans have done the incredible again! One must admit that it's magnificent work."[19]

In reality, Denmark didn't have much choice except to surrender. There was little the country could do to defend itself against the Germans: there were no mountains to which patriots could retreat and strike back at the enemy; the nation was small in size; and the two countries shared a land border, leaving Denmark open to attack by Germany's tanks at any time. Also, the Luftwaffe had destroyed or seriously damaged most of the Danish Army Air Corps' planes when it strafed Værløse Airfield near Copenhagen.

By the end of the day on April 9, 1940, thanks in large part to the element of surprise, German military forces had captured Denmark and all of Norway's chief ports. But they had not captured Norway's King, the Government, nor its gold bullion.

3

Panic and Escape

It seemed almost incomprehensible. In Oslo, crowds of stunned people lined major streets on April 9, watching as German soldiers trooped unhindered into town, some in trucks with rifles at the ready, accompanied by a military band. Several Norwegians chatted and joked with the Germans troops as they marched by.

But the reaction soon changed as rumours spread rapidly that British aircraft were coming to bomb Norway's cities. People tried to escape from the urban areas by any means of transportation they could find. Those who had relatives or summer homes in the countryside headed there. Others flocked to rural hotels. Roadways were clogged with long lines of cars, trucks, and buses. At Oslo's train station, the crush of crowds was so great that some people fainted before they reached the train tracks. Knut Fossum, who was a small boy at the time and whose mother had brought the family to the Oslo station, remembers people running, screaming, and crying as they tried to get on board the trains.[1]

Most came back to their homes after a couple of weeks, but the cities they returned to were drastically different than those they had left. Gunhild Ramm Reistad recalls how her family fled from the city when the Germans invaded, taking refuge in a hotel in the north part of the country for a couple of months. When they returned to Oslo, her father, a lawyer, was arrested by the German authorities and imprisoned for some time. In 1943, the family was evicted from their home and a German official moved in; this misfortune was suffered by many homeowners residing in the outskirts of Oslo. The Ramm family had to live with relatives for nearly half a year until they found another house.[2]

John Stene

Amidst the confusion of April 9, John Magne Stene's first thought was to report to the Reconnaissance Squadron of the Trøndelag Air Unit based at Værnes Air Station near Trondheim.

John was a 25-year-old chemical engineer. He was born in Stavanger, Norway, and was pursuing his career in Mjøndalen, a city southwest of Oslo. He had trained as a pilot while serving in Norway's military service during his university years.

In the fall of 1939, with war imminent, John had been called up for duty by the Army Air Service but sent home again after one month.

Now as German forces attacked Norway, John dressed in warm outdoor clothing, put his uniform in his rucksack, and headed for the train station. On the train, travelling towards Trondheim, he met a Norwegian Army officer who reported that the city was already occupied by the Germans.

John got off at Støren and caught a second train, but when it stopped at Røros for a supper break, he heard that some Norwegian military planes had landed on an ice-covered lake nearby. He felt sure this had to be his squadron, evacuated from Værnes.

One of John's close friends lived in Røros. He lent John some skis and drove him as close as possible to the area where the planes had landed. From there, John skied on in the darkness and encountered an air force acquaintance who had flown in on one of the planes.

John Stene.

PHOTO: NORWEGIAN ARMED FORCES MUSEUM, OSLO, NORWAY.

Together, they made their way to Lake Aursunden where six to eight Army Air Service Fokker biplanes had landed on the lake's frozen surface, including the biplane flown in by Deputy Commander Wilhelm Mohr.

Mohr took command of the squadron after a couple of days. With no air base to return to, the group waited on the frozen lake for 17 days, watching German aircraft regularly soaring above. Petrol-supply problems arose. On two occasions, John flew one of the biplanes across the border to Sweden to pick up medical supplies and deliver them to the Røros hospital. Some pilots carried out reconnaissance missions to map out German ground advances, but, as Wilhelm Mohr wrote in later years, "I would have to characterize the usefulness of our flying as doubtful." It was a frustrating, demoralizing time for the men, sitting and waiting for orders and hearing only sporadic, unreliable information about Norwegian losses in battles elsewhere.[3]

Escape of the King and the treasure

Norway's Foreign Minister, Halvdan Koht, listened to Germany's Minister for Norway, Curt Bräuer, reading out Hitler's demands during their early morning meeting on April 9, 1940.

When Bräuer had finished, and was awaiting a response, Koht consulted with the Norwegian Cabinet members who were conferring in the next room. He returned after a few minutes and declared that Norway would not submit voluntarily to Germany's occupation. Bräuer replied angrily: "Then there will be fighting and nothing can save you." Koht responded: "The fight is already in progress."[4]

Prime Minister Johan Nygaardsvold had been consulting with Cabinet members since around 1:30 a.m. Amid air-raid sirens and blackouts, they tried to piece together what was occurring in their country. About 2:00 a.m., after receiving news of German battleships sailing into Bergen harbour, Prime Minister Nygaardsvold called King Haakon, who was working in his study at the palace, to advise him that the country was now at war. British authorities were contacted and requested to give immediate aid. As Prime Minister Chamberlain later told the British Parliament, Norway was

H.M. King Haakon.

PHOTO: NATIONAL LIBRARY OF NORWAY.
PHOTOGRAPHER: ERNEST RUDE.

assured that full assistance would be coming forthwith and that powerful units of the Royal Navy were at sea.[5]

With Norway now at war, it was important that the King and Storting members not be captured by the enemy. Around 7:30 a.m., King Haakon and a party, including Crown Prince Olav, his wife Crown Princess Märtha, their children Ragnhild, Astrid, and Harald, and most of the members of the Storting, left Oslo by special train for Hamar, a town 130 kilometres to the north.

Shortly after the Royal party left, Norway's treasury began its own journey out of Oslo, barely escaping Hitler's grasp. Early in 1940, Finance Minister Oscar Torp, and Director of the Norges Bank, Nicolai Rygg, had arranged for the evacuation of the country's gold reserves in an emergency. The gold bars and coins held in the Norges Bank's vault had been packed into crates and kegs. Some 300 million kroner's worth of the gold had already been shipped to the United States. About 50 tons of gold, worth 120 million kroner, or 27 million U.S. dollars at the time, remained in the Norges Bank's vault.[6]

Now with the news of German warships in Oslofjord, bank employees loaded the gold onto 26 trucks, which left Oslo and headed north to Lillehammer where the Norges Bank had a large new vault. The last truck left around 1:30 p.m. on April 9, just as German troops marched down Karl Johan Gate. As Oscar Torp later related, a representative of the German government was entering the front door of the Norwegian Finance Ministry as he was leaving by the back door.[7]

The Luftwaffe was bombing Kjeller Air Base when the train carrying King Haakon passed by on its way to Hamar. Train passengers had to disembark and take cover under a railway bridge for about an hour. Crown Princess Märtha later called the bombardment a "most harrowing experience."[8] Other passengers watched as the King, dressed in a grey Norwegian Army uniform, lifted his three-year-old grandson, Prince Harald, to his shoulders and spoke quietly to him, calming Harald and, by extension, those sheltering nearby.[9]

Once reaching Hamar, Crown Princess Märtha and the children made their way to safety in Sweden.

The Storting convened in Hamar during the afternoon of April 9, with only a handful of its members absent. Prime Minister Nygaardsvold offered the resignation of his Labour Party as Norway's Government. Carl Hambro, Leader of the Conservative Party, argued that this wasn't the time for such a change and the Storting unanimously agreed. The Labour Government remained in power, with the addition of three opposition party members to the Cabinet.

Subsequently, the Storting gave Nygaardsvold's Cabinet absolute authority to make all necessary decisions for the security and future of the country, until the Storting could meet again after the war.[10]

Germany's Minister Bräuer contacted the Norwegian leaders at Hamar, asking for a meeting and negotiations: they agreed, presumably because few other options existed. But then events took an alarming twist—news arrived that a German force was heading fast to Hamar. Captain Eberhard Spiller, Air Attaché to Norway, was conducting a daring raid to capture the King and Storting members.

Spiller and Captain Erich Walther, who commanded the companies of paratroopers that had landed at Fornebu in the morning, requisitioned civilian buses and trucks in Oslo. They led a group of the paratroopers north towards Hamar. Fleeing Oslo civilians, whose vehicles were crowding the roadways, slowed Spiller's convoy and, by the time the paratroopers reached Hamar, King Haakon and the Storting members were making their way to Elverum, about 30 kilometres away. Spiller immediately redirected his band to Elverum.

Meanwhile, on the orders of Colonel Otto Ruge, then Inspector General of the Norwegian Infantry, a roadblock had been set up to expedite the King's escape. This roadblock was located at Midtskogen, about six kilometres west of Elverum. It was defended by an informal group of 93 volunteers, composed of Royal Guards trainees, some Norwegian Army officers, and some local gun club members.[11]

In the early morning hours of April 10, Spiller's troops reached the Midtskogen roadblock. A fierce battle raged for over an hour and a half. Captain Spiller was fatally wounded. After Spiller died, Captain Walther and the remaining paratroopers withdrew to Oslo.

And so, King Haakon and the Government members had twice escaped capture, thanks first to Colonel Birger Eriksen at Oscarsborg Fortress, and second to the volunteers at Midtskogen. As the story spread, Norwegian pride stirred. With that pride came a determination to resist the invaders.

Elverum

Vidkun Quisling had not been sitting quietly while the invasion unfolded on April 9. At 7:30 p.m. that day, broadcasting from a local radio station, he declared that a new Nasjonal Samling Government was in power and that he was now Norway's prime minister. Quisling ordered the country's population to cease all resistance activities immediately.

Hitler approved of Quisling's leadership and ordered Minister Bräuer to meet with King Haakon and demand he accept Quisling as prime minister. On April 10, Bräuer travelled to Elverum and met with the King, repeating Germany's earlier

Raid of the Wehrmacht in Norway on April 9, 1940. German troops battle Norwegian soldiers in a village near Lillehammer, Norway. PHOTO: FEDERAL ARCHIVES OF GERMANY, BILD 183-H26353. PHOTOGRAPHER: ERICH BORCHERT.

demands and adding the new one, that the King accept Quisling as his prime minister. When King Haakon told Bräuer that he had to confer with the Cabinet, Bräuer headed back to Olso.

Norway's Government met at an inn located in the nearby village of Nybergsund. In an emotional address, King Haakon stated that it would be a breach of his constitutional duties to appoint Quisling as prime minister, when he knew neither Norway's people nor the Storting had any confidence in Quisling. But if the Government felt it had no choice but to accept the German requirements, he would abdicate.

It didn't take the Government members long to unanimously decide to refuse all Germany's demands, and Minister Bräuer was so advised.

From a small nearby radio station that evening, Norway's Government announced that it had rejected all Germany's demands and called upon the country to resist the invaders.

Colonel Otto Ruge was promoted to major general and named the new commander of the Norwegian Army with the responsibility of overseeing resistance to the invasion.

After a couple of other fruitless attempts to persuade the King to cooperate, the Nazis' tactics changed. Now King Haakon and Government members were to be assassinated.

The Luftwaffe pursued King Haakon and his countrymen from place to place, bombing and strafing their hiding spots. On April 11, German aircraft relentlessly bombed Elverum, leaving it in ruins. That same day, 25 to 30 bombs were dropped on Nybergsund and German planes fired on citizens with machine-guns as they tried to escape from the town's burning buildings. But the King, Crown Prince Olav, Government members, and most villagers were hiding safely in the woods, standing knee-deep in the snow as they watched the town being reduced to rubble.

The King and his party moved northwest towards Åndalsnes.

King Haakon gave an interview to a reporter from *Svenska Dagbladet*, a Swedish newspaper, declaring that he would remain in his country so long as there remained a single inch of Norwegian soil. "My ministers can visit Sweden, my people can evacuate to Sweden. I will have to stay," he said. The

King admitted that he was completely exhausted, not having removed his shoes since April 9 and hardly sleeping. "All civilization seems to have come to an end," he said. "I cannot understand how such terrible things can happen. I can no longer be sure of anything."[12]

Rolf Kristian Hauge

At age 19, Rolf Kristian Hauge from Drammen, Norway, had finished his military training and joined the Navy Air Service. In February 1940, he had volunteered to fight for Finland in its war against the Soviet Union and had been part of a military ski patrol supplying weapons and ammunition to Finnish forces at the front, until the war ended in March of that year.

Rolf, along with some of the Swedish and Norwegian volunteers he'd fought with in Finland, rushed back to Norway after they heard news of the German invasion. The group arrived in the Kongsvinger area a day or two before the German forces reached the city.

That same night, the group gathered ammunition and military equipment and, by horse and buggy, transported it to a spot north of Kongsvinger Fortress to set up a blockade. When the German troops arrived at the blockade, Rolf and his friends fired on them. The German troops were much better armed, for the Norwegians were equipped with only their own rifles. As grenades rained down on the small band of volunteers, they retreated to save their own lives.

The group found a new location for their offensive along the Glomma River, but heard a rumour that Norwegian military forces were fighting Germans to the south. Rolf and two others were sent to look for their countrymen. As the young men skied along, they encountered only abandoned farms and buildings.

Then there was no more snow left for skiing.

Rolf Kristian Hauge.
PHOTO: LITTLE NORWAY MEMORIAL.

They found three old bicycles. Slinging their rifles over their shoulders, they took the bicycles and rode on, until, as they turned a corner, they saw a German tank blocking the roadway. Instantly, German soldiers emerged from the bush and shouted at them.

But when the soldiers looked at the group's Finnish uniforms, they paused. Never having seen Finnish Army uniforms before, the Germans became friendly and inquired as to where the men had come from. Rolf didn't mention that they had just left a group of patriots preparing to attack German soldiers.

The three men were placed under arrest and loaded on buses heading for Olso. There, they were taken to Akershus Fortress and incarcerated with other Norwegian prisoners of war.[13]

4

War Spreads Inland

Germany's military leaders originally believed that Operation Weserübung would be a short campaign. Norway wasn't expected to put up much of a fight.

But to Hitler's surprise, the Norwegians had fought back at Oscarsborg, the King hadn't capitulated, and armed rebel groups had formed and were attacking German troops.

The morning after the invasion, the Royal Navy joined the battle. A flotilla of five British destroyers led by Captain Bernard Warburton-Lee sailed up Ofotfjord to Narvik harbour and took the enemy by surprise. The British succeeded in demolishing two German battleships, the *Wilhelm Heidkamp* and the *Anton Schmidt*, and disabling a third destroyer. Captain Friedrich Bonte, commander of Gruppe 1, was killed on the *Wilhelm Heidkamp*.

By the time the five British destroyers sailed away from Narvik harbour, German warships had reached the area and began to chase Warburton-Lee's flotilla down the fjord. The German ships sank two British destroyers, one being Captain Warburton-Lee's ship. He died in the battle. The other three destroyers escaped, but Narvik remained in German hands.

Britain's Royal Navy returned in strength on April 13, when the battleship *Warspite* led nine destroyers, an aircraft carrier, and other ships in an attack on German destroyers in Narvik's harbour. This time, the British succeeded and seven enemy destroyers were sunk or scuttled. A British reconnaissance aircraft made history by sinking a German U-boat in the fjord; it was the first submarine destroyed by aircraft during World War II.[1]

However, the British didn't immediately land troops in Narvik as naval leaders believed the troops were too exhausted to battle the German troops onshore.[2] Instead, the Royal Navy withdrew from Ofotfjord the next day and sailed home, leaving Narvik under German control.

Now the war in Norway was fully underway. Germany's General Nikolaus von Falkenhorst considered that the most successful strategy for achieving victory in Norway would be for his troops to move speedily into the country's interior, quashing Norwegian defences and linking up with German battalions, which had landed in the coastal cities.

On April 13, Falkenhorst's forces broke out in all directions from Oslo. Divisions moved east, west, and south; two groups moved north towards Trondheim, one travelling through the Østerdal Valley and another heading up the Gudbrandsdal Valley. Using motorized vehicles to transport the soldiers, they made rapid progress.

Norway's Major General Otto Ruge was also considering his options. He determined that any attempt to recapture Oslo would be futile, given the large number of German troops now stationed in and around the city. The best course of action, he believed, would be to retake Trondheim with its excellent harbour and the nearby Værnes Airfield.

In addition, Trondheim had been Norway's capital during Viking times and Ruge believed recapturing the city would lift the Norwegian people's spirits and stiffen their resistance. He contacted British officials, urgently requesting Allied assistance in battling to reclaim Trondheim. Prime Minister Chamberlain himself promised Ruge that help would be coming, as fast as possible and in great strength.[3]

The Allies agreed to land small diversionary forces at the coastal cities of Namsos and Åndalsnes to launch a pincer attack against the German flank at Trondheim. Namsos lies about 131 kilometres north of Trondheim, as the crow flies, and Åndalsnes approximately 196 kilometres south of the latter city. A larger force would land directly in Trondheim fjord.

Ruge decided that his troops would fight in the south of Norway to delay the German advance and so allow the Allies time to win back Trondheim. Once Trondheim was freed, Ruge envisioned a continuing campaign by Allied and Norwegian forces, battling German troops though the rest of the country.

Informal Norwegian Resistance groups continued to surface, setting up blockades and engaging the Germans in fire fights in locations like Hakadal and Bjørgeseter, which temporarily stopped the enemy's march forward.[4]

However, in the face of relentless bombing by the Luftwaffe, and lacking tanks, artillery, or effective air support, the Norwegian forces couldn't prevail for long. Slowly, they were being pushed back to the north.

By April 14, all southeastern Norway had fallen to the Germans.

British troops landed at Namsos between April 14 and 18; French *Chasseurs Alpins*, a mountain infantry, landed on April 19. At Åndalsnes, troops arrived on April 17, 18, and 19.

German forces on the road to Lundebygda in Sør-Aurdal, Norway.

PHOTO: BAGN BYGDESAMLING - VALDRES FOLKEMUSEUM, NORWAY.
REPRO: JOHN ERLING BLAD.

But by now, the Allies had decided it was too dangerous to try and retake Trondheim. Fearing the loss of British battleships to Luftwaffe bombs, they had decided, instead, to fight only from Namsos and Åndalsnes. Winston Churchill told the British Parliament that the landings at Namsos and Åndalsnes had made good progress, and it seemed much easier to capture Trondheim by that method than to incur the heavy cost of a direct attack.[5]

At General Ruge's request, some of the British battalions that landed at Åndalsnes were dispatched inland to prop up the Norwegian forces fighting around Lillehammer.

French, British, and Norwegian forces battled German troops in central Norway, but the constant bombing by the Luftwaffe, and the rapid arrival of German reinforcements, made any offensive extremely difficult for the Allies. Similarly, Namsos and Åndalsnes faced continuous German aircraft attacks, preventing the landing of troops and supplies. Casualties were high.

According to Prime Minister Chamberlain, German control of Norwegian airfields presented the foremost difficulty in defeating the enemy.[6] Fighters flying from airfields in Great Britain didn't have the range to reach Norway, so weren't able to provide effective air cover for Allied troops battling there. To overcome this problem, RAF fighters were being brought to Norway's coast on aircraft carriers; flying from these vessels, the British planes could support Allied troops and attack the Norwegian airfields now held by the enemy.

The RAF intended to go even further, creating an air base near Åndalsnes. However, the area lacked level ground, so the RAF selected Lake Lesjaskog—a frozen body of water located in nearby mountains—for the airfield. Following the RAF's lead, Norway's Army Air Service set up headquarters at Lake Lesjaskog for its few remaining aircraft.[7]

On April 24, the aircraft carrier *Glorious* brought 18 Gloster Gladiator biplanes from RAF Squadron 263 to Norway. The biplanes flew from *Glorious* in a snowstorm and landed on Lake Lesjaskog around 7:00 p.m. that day. The next morning, as the Luftwaffe arrived over the lake, the Gladiator pilots rushed to their planes and tried to start the engines. Due to bitterly cold weather the night before, the carburetors and controls of most Gladiators had frozen, and the pilots were unable to get them off the ground before the Luftwaffe attacked. Thirteen biplanes were destroyed or seriously damaged where they sat, stuck on the frozen landing strip. The air strip had no anti-aircraft guns to defend the Gladiators, as a British ship carrying the weapons and artillery had been sunk by the enemy.[8]

On April 26, the British Government made a significant decision, one that Churchill disagreed with: Allied troops were to be withdrawn from southern and central Norway.

Troop evacuations took place at Namsos and Åndalsnes between April 30 and May 2. Major General Ruge, though very disappointed with the Allies' decision, generously ensured there was transportation available for the troops' withdrawal. British military leaders expressed their gratitude to him and to Norwegian railway staff, who twice drove trains practically to the front lines to collect the troops.[9]

Once the Allied forces had left, Ruge surrendered the Norwegian troops who had been fighting in the central area. These Norwegian soldiers were demobilized; Germany allowed them to return to their homes if they promised not to take part in further hostilities against the occupiers.

In Britain, the failure of the army-navy operations to win any important battles in Norway led to public and political furore. The Allies' Norwegian campaign was viewed as a military catastrophe, with Prime Minister Chamberlain receiving most of the blame.[10] He was unable to form a national government and resigned as prime pinister on May 10, 1940. Winston Churchill became the United Kingdom's new prime minister.

The King and the treasure in hiding

Norway's treasure had remained safely in the Lillehammer bank for 10 days, but the fighting was drawing closer, and it was clear that the gold had to be moved. Norway's Finance Minister Oscar Torp put Fredrik Haslund, Labour Party Secretary, in charge of transporting the gold to Åndalsnes, where it was to be taken by ships to Britain. During the early morning hours of April 19, the gold boxes and barrels, guarded by a handful of Norwegian soldiers, were loaded onto a train heading for Åndalsnes. One of these soldiers happened to be Nordahl Grieg, a well-known poet, novelist, and playwright, who had enlisted in the Norwegian Army to fight the Germans. Grieg would later write poems about his experiences.[11]

The train arrived in Åndalsnes on April 20. But the continual bombing of the city made it too dangerous for the train to simply wait at the station for the British ships to arrive. Haslund arranged for the train to return to a siding near Romsdalshorn, about 10 kilometres away, where mountains protected it from the Luftwaffe. As a result, the country's treasure sat safely in train cars at Romsdalshorn for four days.

On April 24, the *Galatea*, a British cruiser, docked at Åndalsnes. The train returned to the city's harbour and, by dark, soldiers carried some 200 boxes, containing 16 tons of gold, on board the ship. The rest of the gold was to have travelled aboard two other ships, but as the Germans were advancing quickly up the Gudbrandsdal Valley, Haslund decided there wasn't time to wait for them. Instead, the soldiers loaded the remaining boxes on some 25 trucks, which headed for Molde, a city across the fjord from Åndalsnes.

Four German bombers attacked the convoy, but the trucks were able to reach the fjord. The only functioning ferry carried the trucks across the water, two vehicles at a time, in the darkness. It took an agonizing six hours for all the trucks to be ferried across the fjord.

Troops pick through the ruins of Namsos after a German air raid, April 1940.

PHOTO: IMPERIAL WAR MUSEUM, IWM N97.

Back on land, the trucks had to contend with mud and potholes as they travelled along the coastal road. Four of them broke down, but Haslund managed to find three replacements. The heavy cases of gold had to be loaded and unloaded, over and over, as trucks got stuck in ruts and needed to be extracted.

Damaged Messerschmitt Me 110 airplane in the front yard of a house, Junkers Ju 52 aircraft in flight.

PHOTO: FEDERAL ARCHIVES OF GERMANY, BILD 101I-399-0006-19. PHOTOGRAPHER: GÜNTHER PILZ.

Despite the difficult trip and the truck drivers' exhaustion, the treasure arrived safely in Molde on April 29. With the assistance of local citizens, the cases were unloaded from the trucks and hidden in the basement of a factory.

King Haakon, Crown Prince Olav, and the Government members were also in Molde at that time. Just after the group had arrived in the city on April 23, the Luftwaffe began to bomb Molde heavily, and King Haakon and others had to take cover in surrounding woodland time and again. A photograph of the King and Crown Prince standing in the woods beside a silver birch tree would become world famous.[12]

British authorities had offered the King and the Government members sanctuary in England, but the King didn't want to leave Norway; he asked instead to be taken to Tromsø, in the far north, where Norwegian forces were still fighting the Germans.

On April 29, the British cruiser *Glasgow* arrived in Molde to pick up the King and his party, plus the gold. However, while cases of gold were still being loaded on the *Glasgow*, the Luftwaffe attacked the city. The *Glasgow*'s captain decided to set sail right away, as the bombing endangered the King's life, and took along the 23 tons of Norway's treasure that had already been loaded. Over 10 tons remained on the trucks in Molde.

The next step involved taking the last 10 tons of gold by steamer and truck north to Gjemnes, near Kristiansund. From there, five small fishing boats carried the treasure to Tustna. Two larger fishing boats delivered the treasure to Tromsø, arriving there on May 9, 1940.

John Stene

John Stene and his fellow airmen felt encouraged by news of the arrival of British RAF Squadron 263, which was flying Gloster Gladiators from frozen Lake Lesjaskog. The Norwegians were ordered to fly to Lake Lesjaskog and report for duty with the Army Air Service there.

On April 26, John piloted a Fokker biplane south; Niels Evensen[13] was aboard as an observer. It was a beautiful day for flying, with blue sky and white, snow-covered mountains spanning the horizon. They came across a flight of nine German Junker transport aircraft heading towards Trondheim, but the Junkers left the Norwegian plane alone. After an hour and a half of flying, Lake Lesjaskog came into view, and it thrilled John to spot the Gladiators.

King Haakon and Crown Prince Olav shelter in a grove near Molde, during German bombing of the city in April 1940.

PHOTOGRAPHER: PER BRATLAND.

Flying closer, he caught his breath. All the planes were blackened, burned-out wrecks, destroyed by the Luftwaffe the day before. John was staring down at the wreckage in amazement when Niels suddenly banged on his shoulder and shouted, "Damn it all! Remember there is a war on!" Above them, six Luftwaffe Heinkel He 111 bombers circled. John had to land the plane as fast as possible and, in his words, "get the hell out of there."

After landing along the tree-lined shore, John shut off the engine. He and Niels left the plane and scrambled through the snow looking for cover; however, before they had even reached the safety of the trees, one of the Heinkels flew over and dropped its bombs.

The bombs proved to be duds, crashing through the ice without exploding. But the Heinkel passed over again, this time with machine-guns firing. Several bullets hit the Fokker's engine, destroying it.

Once the enemy plane left, John and Niels made their way towards a nearby farm where a friendly family let them stay the night. Later, the young pilots reconnected with their squadron.

On April 29, John flew a Fokker airplane to Nordli in central Norway and reported as a communications pilot for the Norwegian and British forces fighting in the Namsos area. One of John's tasks, while there, was to transport a crate filled with about 500,000 Norwegian kroner (about U.S. $114,000 at that time)[14] to British headquarters for payment of Allied troops fighting near Namsos.

Other Army Air Service planes came to Nordli to carry out reconnaissance duties and support the ground forces' fighting. Around May 1, some 10 Norwegian aircraft flew in from the south, including ones flown by Captain Ole Reistad and Sergeant Per Waaler.

As the Allies were being evacuated from Namsos and Åndalsnes, John's squadron was ordered to leave Nordli and report at Bardufoss Air Station, some 120 kilometres north of Narvik. On the way, the planes landed on ice-covered Langvatnet Lake to spend the night; in the morning, a very strong wind was blowing, and the ice was as smooth as a mirror.

John turned on the plane's engine to warm it up, while he and Niels proceeded to untie the mooring ropes on the wings. A very strong gust of wind blew the plane loose, and the men watched, horrified, as the wind pushed the plane faster and faster across the lake until it finally crashed on the opposite shore.

So, instead of flying to Bardufoss, a trip which would have taken about an hour and a half in a Fokker biplane, John and Niels spent over a week travelling there by boat and bus.[15]

5

The War in the North

With the surrender of much of Norway to the Germans, fighting continued only in the north. There, Norwegian forces were winning victories over enemy forces in the mountains.

Large numbers of Allied troops arrived in the Narvik area in late April and early May 1940, soon outnumbering the enemy: there were approximately 25,000 French, British, Polish, and Norwegian soldiers to some 5,000 German troops.

The Germans still held Narvik and its nearby railway, which transported iron ore from Sweden, but the British intended to try once more to retake the city. The Allies now appreciated that their armies and navies needed effective air cover to succeed in battle, and so awaited the arrival of two squadrons of RAF fighters at Bardufoss Air Station before beginning their attack. Bardufoss had just been expanded to serve as a land base for Allied aircraft; it also served as headquarters for Captain Ole Reistad, freshly appointed by Major General Ruge as deputy head of the Norwegian Army Air Service.

On May 21, RAF Squadron 263, now re-equipped with Gladiators, arrived in Norway on the aircraft carrier *Furious*. On May 26, Squadron 46, flying Hawker Hurricanes, flew in from the *Glorious*. Both squadrons would be based at Bardufoss, along with the one or two serviceable aircraft belonging to the Army Air Service.

The attack on Narvik began early on May 28, when a flotilla of five British destroyers and three cruisers sailed into Ofotfjord and bombarded the harbour area. In the face of heavy machine-gun fire from German soldiers, the flotilla landed French and Norwegian battalions to the north of the city. The Allied troops began to drive the German forces backwards in retreat.

The Hurricanes provided strong air support for the Allies' advance, although the planes were grounded by dense fog at Bardufoss Air Station when they returned for fuel. Coincidently, the Luftwaffe arrived and bombed the British warships in Ofotfjord, causing most to withdraw further out to sea. However, when the fog lifted at Bardufoss, the Hurricanes returned to do battle.

Ground fighting continued all day. Finally, the German commander decided to evacuate his troops from Narvik, and the soldiers slowly withdrew in an easterly direction until they had vacated the city. Around 6:30 p.m. on May 28, Norwegian and French troops entered and occupied Narvik, the first city recaptured from the Germans in World War II.[1] The victory brought joy, pride, and hope to Norwegians—but not for long.

On May 10, Hitler's forces had invaded Belgium, Holland, and France. Two weeks later, the British military leadership decided to transfer all Allied personnel, weaponry, aircraft, and ships from Norway to the new fronts, to fight the Germans there. This included the French, British, and Polish troops, who had doggedly been driving the Germans east from Narvik, and now had them to within a few miles of the Swedish border.

Why would the Allies bother to recapture Narvik on May 28, four days after they'd decided to withdraw all their forces from Norway? Some believe that before the British left Norway, they wished to ensure the destruction of Narvik's harbour facilities to impede future iron ore shipments to Germany. Others suggest that battling for Narvik was meant to mislead the enemy and camouflage the impending withdrawal of Allied forces from the country. Still others argue that the Allies badly needed a victory to raise public morale after their earlier military failures.[2]

Norwegian military leaders weren't told of the planned evacuation until June 1. Their great surprise and disappointment at the news didn't prevent the Norwegians from assisting the Allies' armed forces in their withdrawal.

Troop transport ships arrived off Norway's coast, and between June 4 and 8, about 25,000 Allied troops were evacuated.[3]

On June 8, the RAF's aircraft were preparing to leave Bardufoss Air Station. The Hurricane fighters from Squadron 46, along with Gladiators from Squadron 263, took wing and headed for the aircraft carrier *Glorious*. Landing the Hurricanes on the aircraft carrier was extremely risky, as experts believed the flight deck would be too short for the airplanes. However, the pilots discovered that carrying a small sandbag in the rear of a Hurricane allowed full brakes to be applied, ensuring the airplane would land without tipping over. The Hurricanes, along with the Gladiators, all set down successfully on the *Glorious*.

The next day, German warships sank the *Glorious* and its two accompanying destroyers. The sinking was a naval tragedy: 1,500 died; 40 survived.[4]

Gloster Gladiator.

PHOTO: UNITED KINGDOM GOVERNMENT.

General Carl Fleischer, commander of the Norwegian Army in the north, ordered his forces to surrender to the enemy on June 8. That same day, German troops recaptured Narvik.

King Haakon and the Government members decided to accept Britain's offer of sanctuary; they planned to continue fighting the war against Germany from the United Kingdom. Crown Prince Olav wanted to remain in Norway to encourage

British troops returning from Norway, at Greenock, on board a transport boat.

PHOTO: IMPERIAL WAR MUSEUM, IWM N 383.

the Resistance and show the people that the King and Government hadn't fled the Germans, but this was judged too risky.[5]

The King and Government held their last meeting on Norwegian soil on June 7. King Haakon and Prime Minister Nygaardsvold issued a proclamation, advising that they had been compelled to leave Norway, but they weren't abandoning the fight to regain Norway's independence. On the contrary, they would continue the struggle outside the country's borders.[6]

After this meeting, the King, Crown Prince, Prime Minister, Government members, diplomats, General Fleischer, and some others boarded the British cruiser *Devonshire* and sailed to England. All Norwegian naval ships, merchant marine vessels, and aircraft were ordered to also make their way to the United Kingdom or an Allied port. Around a dozen Royal Norwegian Navy ships escaped to England at that time, along with armed forces personnel, including some 200 from the Army and Navy Air Services.[7]

Only aircraft from the Navy Air Service had the flight range to reach the United Kingdom from northern Norway air bases, and four Heinkel He 115 seaplanes and a captured German Arado Ar 196 succeeded.[8] Another four or five naval aircraft escaped to Finland for internment, as did three planes belonging to the Army Air Service.

King George VI, Prime Minister Churchill, and a cheering crowd met King Haakon and Crown Prince Olav when they arrived at Euston Station in London. Norway's Government-in-exile established itself at Kingston House in the Knightsbridge area of London.

Before the Nygaardsvold Government departed, it had given Major General Ruge command of all military forces in north Norway. On June 10, 1940, hostilities ceased. Ruge negotiated a cease-fire with the Germans, by which the Norwegian Army capitulated, and the Germans allowed members of the armed forces to return home. Ruge would spend the rest of the war in Germany as a prisoner of war.[9]

And what of the country's gold? On May 23, the gold bullion sailed safely from Tromsø to England aboard the British cruiser *Enterprise*. On June 15, the Norwegian freighter *Bomma* brought the first load of gold to New York; 13 similar transports followed. Some two and one-half tons of the gold were deposited with the Federal Reserve Bank in New York and seven and one-half tons with the Bank of Canada in Ottawa. Both banks returned the treasure to Norway in 1987, when its estimated value totalled U.S. $156 million.[10]

The war in Norway was over.

John Stene

On June 6, John heard credible rumours that all British military forces were pulling out of Norway.

As the war in Norway drew to its end, John was ordered to fly one of the last remaining Norwegian aircraft to Finland to be interned for the rest of the war. On June 8, he delivered a Fokker biplane to Petsamo, Finland. A couple of weeks later, after obtaining clearances to return to Norway, he sailed by ship to Stockholm and took a train to Trondheim. To John, it felt very strange to be back in his own country, but one now occupied by the enemy.[11]

6

Merchant Marine

King Haakon didn't arrive in England empty-handed.

Although Norway was a small nation with a population of only 3,000,000 people, it possessed the third-largest merchant marine in the world.

After the Government-in-exile had ordered all merchant marine vessels then at sea, totalling some 1,024,[1] to proceed to Allied ports rather than to return to Norway, Quisling gave a contradictory order. He demanded that all Norwegian ships return to Norway, Germany, or a neutral country and await further instructions. Not one of Norway's merchant marine ships obeyed Quisling's order.

After King Haakon and the Government-in-exile reached England, they offered the services of the Norwegian merchant marine to the Allies. Working with Norwegian ship owners, Government-in-exile officials organized a company called the Norwegian Shipping and Trade Mission or "Nortraship." Through Nortraship, the merchant marine made an enormous contribution to the war effort. Until 1942, half the fuel and one-third of the all other supplies delivered to Britain arrived on Norwegian merchant marine vessels.[2]

Camp Norway, Lunenburg, Nova Scotia

Many ships from the Norwegian merchant marine sailed into Halifax, Nova Scotia, shortly after April 9, 1940. This included Norway's whaling fleet with some 2,000 sailors, who had been heading home from the Antarctic whaling grounds when Germany invaded.

Nortraship set up an office in Halifax, as did the Royal Norwegian Navy. The Norwegians also established a hospital, a seamen's church, and a seamen's club. In Lunenburg, about 90 kilometres southwest of Halifax, the Royal Norwegian Navy set up a naval and military training school called "Camp Norway," where merchant marine sailors received training as gunners before returning to sea.

In addition, the Norwegian Army operated a base in Lunenburg from the spring of 1942 until the summer of 1943.

Serving on a merchant marine ship during wartime could be a very dangerous occupation. Over the course of the war, the Norwegian merchant marine lost 570 vessels and 4,000 seamen.[3] Aware that their ships might be attacked by German torpedoes at any time, sailors took precautions when passing through hazardous waters. As Radio Officer Kai Pedersen described in his book, *Viking Ships Had No Lifeboats and Other Stories*,[4] he and other sailors slept on the deck every night, using lifebelts as pillows, ready to head for the lifeboats if their ship was hit by enemy fire.

Reorganizing Norway's military

King Haakon, the Government-in-exile, and Norwegian military leaders wanted Norway to do more than contribute its merchant marine to the war effort, huge though that contribution was. They visualized a Norwegian military that would battle alongside the Allied forces and, when the time came, lead Britain and the others in the recapture of Norway. The question was, what form should Norway's military contribution take?

The British Government suggested that Norway's pilots, based in Britain, be allowed to join RAF units, but Captain Bjarne Øen of the Norwegian Army Air Service argued against this. Echoing a view formerly expressed by Major General Ruge, Øen wrote: "Seen from outside, nothing much will be gained by letting our people join individually within the Allied forces. It must be Norwegian units that can show our colours, and that will stand identifiable when victory comes and settlement takes place."[5] The idea of creating distinctive Norwegian units within the Allied air force became the goal of the Government-in-exile.

Britain and the Government-in-exile reached an agreement during the summer of 1940, later formalized on May 28, 1941, resolving the matter.[6] All Norwegian Armed Forces in the United Kingdom would be used for the purpose of either defending the United Kingdom or regaining Norway. They would be organized under British Command, as the "Armed Forces of the Kingdom of Norway allied with the United Kingdom." Norway's Government-in-exile was to bear all the costs of equipping, paying, and maintaining its own armed forces.

As requested by the Government-in-exile, the Agreement also specified that certain Norwegian Air Services units, manned by Norwegians and flying under the Norwegian flag, would be attached to the RAF and be under British operational control.

As a result of the Agreement, those Royal Norwegian Navy ships that had followed King Haakon to Britain came under the Royal Navy's command and began carrying out patrols in the Arctic. The Norwegian Army also established a training camp in Scotland to prepare recruits for their service in the eventual recapturing of Norway.

However, it wasn't so easy for Norway's Army and Navy Air Services to find a home. Even before Norway surrendered, military representatives, including Captain Øen, had discussed plans with the British authorities for training Norwegian military pilots at a location somewhere in the United Kingdom. But following the German invasion of Holland, Belgium, and France, the RAF required all the United Kingdom's training fields for its own use; there were no surplus airfields to offer to the Norwegians. The British did propose that Norwegian air crews be trained as volunteer reservists, who would later serve in RAF squadrons in northern Norway; however, after Norway surrendered to Germany, this idea was abandoned.

Norwegian representatives had also been considering the south of France as an alternative location for pilot training until that country fell to German forces.

Canada became a likely choice for several reasons, the chief one being that the American aircraft ordered by Norway before the war, but not delivered prior to April 9, 1940, could be speedily and easily provided to a Canadian location. Also, the country was located at a distance from the war zone, and provided protection for the young recruits as they learned their skills. Finally, United Kingdom air crews were already being trained in Canada under the British Commonwealth Air Training Plan.

The British Commonwealth Air Training Plan

At the beginning of World War II, Britain looked to the Commonwealth to assist it in training air crews and ground crews for service with the RAF. The lack of space for such schools in the United Kingdom created a real difficulty, one that could be solved by moving the training elsewhere.

Britain approached Canada, Australia, and New Zealand to join in a plan to develop pilots, mechanics, gunners, observers, and other air force personnel for eventual fighting overseas with the RAF. Canada was chosen to be the primary site for the British Commonwealth Air Training Plan (BCATP) for a number of reasons. In comparison to Australia and New Zealand, Canada had a greater industrial capacity to manufacture planes; its nearness to Britain made transportation of personnel and equipment easier; and it offered better access to American aircraft manufacturers.[7]

All four countries signed an agreement formalizing arrangements for the BCATP in December 1939.

Recruits from Australia, New Zealand, and the United Kingdom took early flight training through expanded programs in their home countries and then travelled to Canada for advanced training. Recruits from Canada took all levels of training at home.

Canada administered the BCATP. As Prime Minister Mackenzie King told the Canadian Parliament in May 1940, this involved, in part, operating 3 Initial Training Schools, 26 Elementary Flying Training Schools, 16 Service Flying Training Schools, 10 Air Observer Schools, 10 Bombing and Gunnery Schools, 2 Air Navigation Schools, and 4 Wireless Schools.[8] Some 60 new airports/airfields would need to be built to implement the BCATP and another 20 expanded.[9]

The length of training varied with the occupation: for pilots, it took about 28 weeks; for air observers, 26 weeks; and for air gunners, 32 weeks.[10] The BCATP had an ambitious goal: 50,000 air crew were to be trained each year.

To get things moving, Canada contracted with local flying clubs and airlines to operate a number of Elementary Flying Training Schools and all of the Air Observer Schools. In addition, the Toronto Flying Club ran the No. 1 Initial Training School, which opened on April 29, 1940, at Malton Airport.[11]

Arne Skretteberg and Svend Clement

Arne Skretteberg was one of the Norwegian sailors who had been at sea on a whaling ship when the Germans invaded Norway. The 23-year-old had been a whaler for five years. He was six feet tall, very powerfully built, and a gifted athlete. Before going to sea, Arne had set track and field records in Norway.

After receiving orders from Norway's Government to sail to an Allied harbour, the captain of the whaling ship considered heading to Iceland. However, he determined that this would be too dangerous given the number of German submarine strikes along that route and instead, sailed to Lunenberg, Nova Scotia.

Once in Lunenberg, Arne had to decide whether to continue to work with the merchant marine or join the Norwegian military forces regrouping in Canada. He chose to join Norway's Army Air Service.[12]

Svend Clement, from Copenhagen, had been serving as a ship's

RIGHT

Svend Clement.

PHOTO: LITTLE NORWAY MEMORIAL.

FAR RIGHT

Arne Skretteberg, front row, second from right, wearing light-coloured uniform. PHOTO: MICHAEL SKRETTEBERG COLLECTION.

BOTTOM RIGHT

SS *Europa*.

PHOTO: LITTLE NORWAY MEMORIAL.

steward on the EUROPA, *a Danish cargo and passenger vessel, when the Germans had invaded Denmark. The* EUROPA *immediately sailed to New York City and then to Halifax, where it was placed at the Allies' disposal.*

The Canadian government took over the ship, which then operated as part of Canada's merchant marine, carrying supplies, troops, and war material across the Atlantic.

When in Nova Scotia, Svend Clement, like Arne Skretteberg, chose to leave the merchant marine and sign up with the Norwegian Army Air Service.[13]

WRF-202

7

A Canadian Home for Norwegian Air Services

Negotiations to establish an air training centre in Canada for the exiled Norwegian airmen began in early June 1940. Norway's Consul General, Daniel Steen, met with Canada's Under-Secretary of State for External Affairs, Oscar Skelton, and military officials. The discussions soon led to a crucial question: where should the Norwegian Army and Navy Air Services make their home? A quick decision was needed, and the choice was narrowed down to Dartmouth, Nova Scotia, or Toronto, Ontario.[1]

Captain B. Fredrik Motzfeldt, Assistant Military Attaché to the Royal Norwegian Legation in Washington, and Commander Hjalmar Riiser-Larsen, Naval Attaché to that Legation, travelled to Toronto on June 30, 1940, to inspect the Toronto Island Airport as a possible site for the Air Services' training camp.

A *Toronto Daily Star* reporter described Riiser-Larsen as "huge, good-looking, dark-haired," and noted he had been second-in-command on Roald Amundsen's attempted flight over the North Pole in 1925.[2] When their plane was forced down in the Arctic, its occupants had spent 26 days on an ice shelf and nearly died before escaping. In 1926, Riiser-Larsen had joined Amundsen in another attempt to fly over the North Pole, this time succeeding. He had also taken part in expeditions to Antarctica. Riiser-Larsen was about to be appointed commander of the Navy Air Service.

Toronto Island Airport

Toronto Island Airport was owned and operated by the Toronto Harbour Commissioners, a joint federal-municipal agency established by the Canadian government in 1911. The Toronto Harbour Commissioners were given responsibility for Toronto Harbour and the city's waterfront.

Motzfeldt and Riiser-Larsen met with Airport Manager W.J.E. (Edward) Johnston, and examined the facilities. The Norwegians reported back to their superiors that the airport seemed to give the best promise of a combined land-and-sea base for a moderate number of planes.[3]

Another individual favouring Toronto was Colonel Bernt Balchen,[4] a Norwegian World War I veteran and, like Riiser-Larsen, a famous polar aviator. He had flown across the Atlantic with Admiral Richard Byrd in 1927. In 1929, he joined Byrd again when Byrd made the first flight over the South Pole.

Balchen was a member of Norway's Navy Air Service and

OPPOSITE Looking northwest across Toronto Island Airport and Bathurst Quay, April 1941. Little Norway lies on the north side of the Western Channel.

PHOTO: PORTSTORONTO ARCHIVES, PC 15/3/848. ORIGINAL, LIBRARY AND ARCHIVES CANADA.

Bernt Balchen.
PHOTO: SAS SCANDINAVIAN AIRLINES.

held dual Norwegian-American citizenship. Before the war, he had been negotiating with the United States on behalf of Norway for the purchase of aircraft and weapons. Balchen also had contacts in the Canadian military. In late April 1940, he had travelled to Ottawa to meet with government and armed forces representatives, seeking permission to haul nearly 100 American airplanes across the Canadian border for their further transport to Norway. These aircraft had been ordered from American suppliers before the war.[5]

With Balchen, Riiser-Larsen, and Motzfeldt favouring Toronto, Norway's Government-in-exile settled on the city as the site for the new air-training camp.

The British government and Canada's Department of National Defence for Air soon approved and authorized a Toronto training school for Norwegian air and ground crews.[6] The Canadian government exempted the Norwegians from compliance with certain air regulations relating to registration and marking of planes, and licensing of personnel, which would have otherwise applied to them as "foreign military aircraft flying over or alighting in Canada."[7]

The Harbour Commissioners told the Norwegians that they would be given the use of Toronto Island Airport on the same terms and conditions as applied to the Royal Canadian Air Force (RCAF). That meant Norway would have to pay for hangar storage at a specified rate and buy all the aircraft gas and oil needed at Toronto Island Airport from the Harbour Commissioners. The Norwegians advised the Harbour Commissioners that 36 Curtiss P-36 Hawks, 36 Douglas 8A-5 attack bombers, and 24 Northrop scout and bombing seaplanes, ordered from United States manufacturers before the war, would be delivered to Toronto Island Airport over time. The Norwegians also speculated that housing could be required for some 200 men.[8]

Both the Norwegians and the Harbour Commissioners were content with their preliminary understanding of the rental arrangement.

The City of Toronto's Board of Control unanimously approved the Norwegians' use of Toronto Island Airport.

On July 4, 1940, Toronto newspapers excitedly reported that Norwegians would learn to fly in the city. Colonel Oscar Klingenberg, Military Attaché to the Royal Norwegian Legation in Washington, was interviewed and one of his comments, as reported by a local paper, would haunt Norway's Government-in-exile for the duration of the war.

In the article, Colonel Klingenberg began by explaining that he was arranging to bring 125 Norwegian flyers to Toronto to take a post-graduate course on the latest types of American planes. He then stated that the Toronto Island Airport would be used for air training at no cost to the Norwegian government, except for incidental expenses. Klingenberg praised Canada's Under-Secretary of State for External Affairs and Toronto's mayor, with whom the Norwegians had been negotiating, and said, "they have given us everything we asked for, and more."[9]

These observations may have contributed to a widespread public belief that Canadian taxpayers would have to pay for the cost of construction and operation of the Norwegian training camp, at a time when Canada had barely left the Depression and had war obligations of its own.

But that wasn't true. Norway's Government-in-exile bore the full cost of its armed forces' expenses, thanks to the income flowing to it from the Allies' use of Norway's merchant marine. Norwegian spokesmen made this point over and over in later public communications, but the mistaken belief that Canadians were paying for Norwegian air training continued to percolate throughout the war.[10]

Agreement with the Toronto Harbour Commissioners

Two flight lieutenants from the Army Air Service, Olav Bakke Stene and Odd Bull, and one from the Navy Air Service, Kaare S. Kjos, arrived in Toronto around July 8 to supervise delivery of the new American airplanes and the preparations

Toronto Island Airport, 1940, with Toronto Harbour beyond.

PHOTO: TORONTO PUBLIC LIBRARY.

for the training camp.[11] Olav Stene was John Stene's older brother and had left Norway around the same time as the King and Government-in-exile.

Olav Stene and Odd Bull had authority to negotiate a formal agreement with the Harbour Commissioners for Norway's use of the Toronto Island Airport. On September 13, 1940, a written contract between Norway and the Harbour Commissioners was finalized.[12]

Here is a closer look at this Agreement.

The Harbour Commissioners granted Norway permission to use Toronto Island Airport as a base for land planes and seaplanes, and for pilot training. However, Norway had to share use of the airport with others who had contractual rights to the property, such as Island Airport Air Training and Transport Ltd., a company offering private flying lessons. In addition, British or Commonwealth air forces, like the RCAF, had priority over the Norwegians in using the airport.

The Agreement gave the Harbour Commissioners' airport manager power to coordinate the movements of the various aircraft using the airport.

Other provisions dealt with space rental. Norway agreed to pay the Harbour Commissioners a monthly charge for mooring seaplanes and storing aircraft, based on a plane's wing span. Office space could be rented in the administration building at the airport, at the rate of $1.50 per square foot per year.

Norway was required to purchase all its gasoline and oil at the Toronto Island Airport, at the Harbour Commissioners' current retail prices, as set from time to time.

Personnel who travelled by ferry to and from Toronto Island had to buy tickets, and these could be purchased at a special rate of 28 tickets for $1.00.

The Norwegians needed a mechanical overhaul building and agreed to build it at their own expense, with a small contribution from the Harbour Commissioners. No rent would be charged for the use of this building but at the end of the war, or if the Agreement terminated before then, the building would belong to the Harbour Commissioners.

The Harbour Commissioners agreed to make land available on the mainland, west of Bathurst Street and north of the Western Channel, for the Norwegians to build accommodation for personnel. No rent would be charged for use of this land, but at the war's end, or if the agreement terminated before then, Norway had to offer to sell the buildings to the Harbour Commissioners. If no bargain could be reached, Norway had to remove the buildings and restore the land to its original condition.

The Agreement also contained other provisions, such as obliging the Norwegians to obey Canadian aviation laws and make monthly reports to Canada's Department of Transport about the number of aircraft at their base.

Major General William Steffens, a commander of Norway's west coast defences during the war with Germany, had been appointed head of the Norwegian Military Mission in Canada, and signed the Agreement on behalf of King Haakon and Norway's Government-in-exile.

With that, the Army and Navy Air Services had a new home in Toronto.

Who should be trained?

Interestingly, there were some different views among the Norwegian representatives as to who should be entitled to take the air-training courses. Bernt Balchen told interviewers in July 1940 that he would like to take good American pilots who were over the age limit of the RAF or RCAF, bring them to Canada to train on the United States' modern fighter planes, which Norway had ordered before the war, then send them to Norway to do battle.[13]

Others thought that the schools should only instruct those who were already pilots. Flight lieutenants Olav Stene and Odd Bull told the *Toronto Daily Star* on July 20, 1940, that only Norwegians with previous flying experience would be trained. Flight Lieutenant Stene said 70 pilots, some already members of the Norwegian Air Services, were in England waiting to come to Canada, and their number would be increased by Norwegians in Canada and the United States who had flying experience. All would be given a refresher course, including blind flying, bombing, gunnery, and wireless operation. Ground and air crews would be chosen for training from among Norwegian sailors whose ships moored in Allied ports after the German invasion, Stene said.[14]

Per Conradi Hansen, a former reporter and subsequent member of the Royal Norwegian Air Force (RNAF), wrote in his book, *Little Norway: A Message of Liberty to the Hills of Home*:

> It is quite obvious that they were blind to the fact that the training of pilots and other aircrew was comparatively quick and simple. They were thinking in pre-war terms of well-turned out officers who had been carefully selected with regards to education, physical health and fitness, family background (some income, outside the Air Force, was almost a necessity), and general good appearance.[15]

However, pilots, navigators, and other future members of the Norwegian air-training camp were already making their way to Canada on two small steamships, the *Iris* and *Lyra*, sailing as part of a convoy travelling from Glasgow, Scotland, to Montreal, Quebec. Aboard were some 120 Norwegian recruits and officers, including many skilled members of Norway's Air Services, and they would form the core of the first wave of airmen. One of these was Harald Olafson.

Harald Olafson

Harald Olafson, from the Østerdalen area of Norway, had been working as a civil engineer in Buenos Aires, Argentina, when Germany attacked Norway. As a second lieutenant in the Army Air Service, he received a telegram ordering him to report for duty as soon as possible. Harald took the first available ship bound for Nor-

Harald Olafson.

PHOTO: OLAFSON FAMILY COLLECTION.

way. The ship was heading for Bergen, loaded with food supplies for the Allied forces.

Nortraship sent a message to the ship at sea, ordering its captain to proceed to Sierra Leone, Africa, and await a convoy of vessels coming from Cape Town. The captain complied, and around May 29, Harald's vessel joined a convoy sailing for the United Kingdom. After arriving safely in the English Channel, the ship travelled to Ireland, where its food cargo was unloaded. From there, Harald made his way to London, England, and reported for duty with the Norwegian authorities.

After a few days, orders came for the Army and Navy Air Services personnel to move to RAF Dumfries Air Base in Scotland, where they remained for a month and a half.

Lieutenant Wilhelm Mohr visited the Norwegian Army Camp, also at Dumfries, where he met about 60 former seamen. Many of these men were whalers. Mohr recruited them to travel to Canada and join the Norwegian Air Services training there; most would become aircraft mechanics.[16]

At long last, Harald and the other officers, along with the new recruits from Dumfries, boarded two steamers, the IRIS *and the* LYRA, *which must have seemed tiny among the other vessels in the convoy of over 100 ships gathered in Glasgow harbour. Before the convoy left for North America, German planes attacked the ships, but there was no serious damage. The convoy proceeded on its course across the Atlantic, once escaping an attack by a German submarine, before splitting up as it reached various ports of call. The* IRIS *and* LYRA *sailed into Montreal harbour around July 2, 1940. From there, Harald and many others took a train to Toronto.*[17]

8

Little Norway in Toronto

The Army and Navy Air Services' new training camp, Little Norway, lay at the foot of Bathurst Street in Toronto, directly across the Western Channel from Toronto Island Airport. It encompassed some 6.7 acres or 2.7 hectares.

Two commanding officers shared leadership duties at Little Norway: Major Bjarne Øen for the Army Air Service and Captain Hjalmar Riiser-Larsen for the Navy Air Service.

While the camp's barracks were under construction, Little Norway's personnel lived in temporary accommodations. Most of the officers stayed at the Royal York Hotel, which also served as headquarters for both Air Services.

To house the large group of recruits who had come to Canada on the *Iris* and *Lyra*, the Norwegians rented nearly an entire floor of the Prince George Hotel, located at the corner of York and King Streets. The Air Services also leased the vacant Lakeside Home for Little Children on Centre Island, which was owned by the Toronto Hospital for Sick Children and had been used in the past as a summer branch of that hospital. In addition, some men remained living on the *Iris*, now docked at Toronto harbour.

On September 1, the Army and Navy Air Services established permanent headquarters on the fourth floor of a building at 341 Church Street. The Air Services shared the 4,500-square-foot space. Each service had its own individual branch for purchasing, training, equipment, and pay and accounts, but shared medical, recruiting, and public relations functions.

Training begins

Although Little Norway's personnel were keen to begin air training, the camp had no aircraft of its own. The Toronto Flying Club, based at Malton Airport, came to the rescue. In late August 1940, 20 seasoned pilots, most of whom had arrived on the *Iris* and *Lyra*, attended a month-long refresher course given at the club.[1] Members of this group became flight instructors when Little Norway received its training aircraft.

During the summer of 1940, the Norwegians ordered 36 Fairchild PT-19 aircraft from the United States to serve as basic training planes for the Army Air Service; the first one arrived at Toronto Island Airport on August 23, 1940.

The Fairchild PT-19 was a monoplane with a bright blue body and yellow wings; the Norwegians added stripes to the tail and wings in their national colours of red, white, and indigo blue. The aircraft had an open cockpit with two seats, one at the front for the student pilot, and one behind for the instructor. As it was a training plane, the Fairchild offered dual controls and instrument panels. The plane, which cost about $12,000 to purchase, could reach a maximum speed of some 160 kilometres per hour.[2]

Camp Little Norway, Toronto. Moving clock-wise from the ferry slip, there is a grouping of four buildings. The first, running north-south, served as officers' quarters, as did a second building to its south, bordering the Western Channel. The other two buildings in the grouping housed non-commissioned (NCO) officers. Two H-style buildings follow: the one to the north held the NCO officers' mess and the other, the officers' mess. The hospital lay to the left of these structures, abutting the Western Channel.

North of the hospital were two H-style barracks for airmen, with showers and lavatories located in smaller connecting buildings. Further to the north were the boilers, Radio City, and Haraldshallen, which served as a gymnasium. Continuing to move in a clock-wise direction, a long building situated parallel to Maple Leaf Stadium contained the airmen's mess, the kitchen, and the canteen. The next served as the camp school. Administration offices could be found in a building further to the east. Other camp structures included the guard house, garages, and buildings housing equipment. A parade ground lay in the middle of the camp.

PHOTO: JON TVEDTE, *FLYVNINGEN. DET MODERNE EVENTYR*, OSLO, 1958.

To help meet the demand for aircraft, the Norwegians temporarily rented planes from Leavens Brothers Air Services Ltd., a company that operated a flying school at Toronto Island Airport.[3]

Deliveries of more aircraft soon followed. By October 1, Little Norway's Army Air Service had eight Fairchild PT-19s and one Waco E biplane at its disposal; the Navy Air Service possessed two Stinson Reliant SR 9C seaplanes.[4] The Douglas, Curtiss and Northrop aircraft ordered before the war soon would soon begin to arrive; the goal was for these planes to accompany the fully trained air crew and ground crew when they transferred to the United Kingdom to join the Norwegian squadrons of the RAF.

The camp

The camp was composed of some seventeen buildings, including barracks, messes, a ground school, and a hospital. The buildings encircled a grassy parade ground, where every day, trainees lined up for inspection before and after work. A steel barb-wire fence enclosed the entire camp, except for a large gate for vehicles and a small one for personnel.[5]

New recruit, Harald Jensen of Larvik, Norway, later wrote, "As camps go, it was exceptionally pleasant.... The mess was sparkling clean, and the food was excellent."[6]

Most of Little Norway's buildings were grey with bright red trim. Gingham curtains in the windows warmed the atmosphere: some 1,500 pairs of curtains had been sewn by local Toronto Norwegians to brighten up the surroundings.[7] Framed photos of King Haakon and the Royal Family hung in many locations. A reporter wrote:

> Of all the military camps that I have visited, Little Norway, the main headquarters in Toronto of the Norwegian air force in Canada, certainly takes the prize for orderliness, cleanliness, military snap, and strangely enough, cheerfulness.[8]

Opening day

Rain poured down on some 500 guests gathered at noon on November 10, 1940, for Little Norway's opening ceremonies. Canadian-born actress Mary Pickford watched the event along with her husband, actor and band leader Buddy Rogers. Bernt Balchen attended and so did Lars Christensen, a Norwegian whaling magnate, whose wife and daughter were said to be "dripping with magnificent furs."[9]

Official guests included Norway's Consul General, Daniel Steen; Commander in Chief of the Norwegian Armed Forces in Canada, General W. Steffens; Canada's Air Vice-Marshal, Lloyd S. Breadner; RCAF Air Commodore G.E. Brookes; Ontario's Lieutenant Governor Albert Matthews; Toronto's Mayor Ralph Day; and a host of officials from Ottawa, Montreal, New York, and Washington.

Units of men from Norway's Army and Navy Air Services, and from the RCAF, stood outside in chilly weather as the ceremony began. General Steffens inspected the troops. On two identical flagpoles, Norway's flag and the RCAF ensign were slowly raised, followed by the playing of "O Canada" and then a patriotic Norwegian song, "*Ja, Vi Elsker Dette Landet*" ("Yes, We Love this Country"). General Steffens spoke of the German invasion, as the audience listened solemnly:

A Fairchild training plane in flight. The Norwegian flag's colours, red, white, and blue, are painted on the aircraft's wings and tail.

PHOTO: ROLF KOLLING COLLECTION.

> We Norwegians who are present have no more homes ... our dear ones whom we left behind us in Norway live under tyrants. But we stand here today resolved to play our part in liberating Norway and we know that this will happen.... But until victory is achieved, and we again can return to our own country, our own homes and our dear ones, all our strength must be directed against breaking the chains of Nazi enslavement and restoring freedom to the world once more.[10]

Exercises at Recruit School, Lakeside Home, Centre Island, Toronto.

PHOTO: GRAVENHURST ARCHIVES.

After other speeches, Reverend B. Overland, a minister from Montreal, led a prayer and hymn in Norwegian, bringing the official ceremony to an end.

Overhead, 11 Norwegian aircraft flew in tight formation. Roaring, the planes dipped to within 25 feet of the new barracks, thrilling the crowd.

The guests toured the camp and ate lunch in the officers' mess. Here is how one happy diner described the food:

> Long tables groaned with huge open-faced sandwiches of every fish you could imagine, prepared by Norwegian chefs; [there were] enormous cakes smothered in whipped cream and gay with Norwegian colours.[11]

An *Evening Telegram* reporter noted large platters of pickled herring, anchovies, spice cakes, and all sorts of smorgasbord concoctions served on generous wedges of rye bread. She claimed never to have seen anything disappear as fast as a large cream cake or *Blotkake*: several of the young officers said they hadn't tasted any since happier days at home.[12]

Exiles

In 2012, Wilhelm Mohr told a *Gravenhurst Banner* reporter what it was like for the Norwegian military personnel to arrive in Toronto. The former lieutenant general was 95 years old when interviewed. He said, "We weren't all happy going to Canada at the time." The men were keen to be in Norway defending their homeland, rather than across the ocean, far removed from the theatre of war. However, the Canadian people and military personnel accepted the Norwegians with open arms. "The support was amazing, amazing, amazing. We were made to feel that we weren't that far from home."[13]

One pilot recalled arriving in the camp after long travels, and seeing old friends in new blue uniforms. "They had new airplanes and new buildings and it was an amazing experience ... all the buoyancy. The boys were flying at night in these new airplanes which looked beautiful to me."[14]

School

When a civilian arrived at Little Norway to join the Army or Navy Air Services, the next week or so would be spent undergoing health checks and security screening. Once accepted into one of the Air Services, a recruit received a service number and a *Vernepliktsbok*, or Record of Service Book, in which to record postings, courses attended, promotions, and similar matters.

Recruit School followed: four weeks of drilling, marching, calisthenics, studying military procedure, and learning the use of arms. The first session, held at Lakeside Home, began in October 1940, before Little Norway's official opening.

After Recruit School, instructors determined where a recruit should be placed in the air-training program. They considered the individual's abilities, education, effort, maturity, stability, and behaviour, both on and off duty, in making this decision.[15] Not everyone could become a pilot, although almost all the recruits wished to do so. The Norwegian Air Services needed people in different positions: it was said that for every man in the air, some 10 were needed on the ground.[16] Mechanics, service police, cooks, drivers, medics, administrators, and many other types of personnel were essential to keep the camp operating.

Sergeant Arne Skretteberg had intended to be an aircraft mechanic, but the camp's commander was aware that Skretteberg had set records in track and field in Norway, and asked him to become Little Norway's physical training instructor instead. Skretteberg agreed.[17]

Young Norwegian trainees who wished to become navigators, air observers, wireless operators, gunners, flight engineers, and air bombers attended schools related to their individual areas of expertise. On graduation, air observers, navigators, and air bombers received a one-wing badge and a promotion to sergeant.

Lakeside Home for Little Children on Centre Island, formerly used during the summer for patients from the Toronto Hospital for Sick Children. In 1940, the building was vacant and the Norwegian Air Services rented it to serve as a barracks and Recruit School. PHOTO: JON TVEDTE, *FLYVNINGEN. DET MODERNE EVENTYR*, OSLO, 1958.

Ground crew, such as mechanics, took classes at Little Norway plus additional specialized courses at other locations. Mechanics School offered a 26-week course, focusing on engines and air frames.[18]

Little Norway strongly promoted education for personnel working in all types of jobs, including those in administration.

The students worked hard at their lessons; keen to finish their training and return overseas to fight the Nazis, they often studied until late in the night. "The lads training in Toronto now," one instructor remarked, "don't have to be urged or driven. They're grabbing every atom of training they can get."[19]

There was the odd mishap. A trainee-mechanic demonstrating switches on a Stinson seaplane in the hangar at Toronto Island Airport accidently released a magnesium parachute flare. The flare dropped out of the bottom of the aircraft, burst into a hot flame, and began streaking across the floor. It hit the wall, bounced back, and stopped under the airplane from which it had originated, exploding under its left pontoon and burning with terrific heat for about 40 or 50 seconds until extinguished. The trainee received a reprimand.[20]

Rolf Kolling

Rolf Kolling was born in Oslo, but his family moved to Victoria, British Columbia, around 1909, where they stayed for a decade before returning to Norway.

In the fall of 1939, 18-year-old Rolf was working as a seaman in the engine room of the HAV*, a brand-new cargo ship, which had sailed from Norway to Canada. On April 9, 1940, when Hitler invaded Norway, the* HAV *had just left Port Alberni, British Columbia, and was heading for South Africa with a cargo of lumber and other goods.*

Canadian authorities ordered the ship to return, and the vessel spent a week in Victoria, British Columbia, before resuming its voyage. While the ship was moored, Rolf visited with some old family friends living in Victoria. Afterwards, he asked the Captain for permission to remain in Canada, but the Captain refused, as he felt responsible for the young crew member.

By the time the boat docked in South Africa, Rolf had heard about Norway's Air Services' training camp at Little Norway. He asked the Captain for permission to return to Canada, but the Captain wouldn't agree.

Rolf Kolling.
PHOTO: ROLF KOLLING COLLECTION.

The HAV *continued to Australia. In Melbourne, Rolf quietly left the ship and struck off on his own. He found employment with another vessel, which was heading to New York. After landing in that city, he travelled to Toronto. Rolf arrived at Little Norway in May 1941 and joined the Army Air Service.*

One year later, the HAV *struck a mine while sailing in the Mediterranean Sea near Port Said, Egypt. Among the dead was a mechanic working in the engine room—in the very seat Rolf had always occupied.*[21]

Pilot Training

While training and courses changed from time to time during the war years, the basic route a trainee followed to become a pilot contained these steps: Recruit School, Initial Training School, Elementary Flying Training School, Service Flying Training School, and Operational Training.

Initial Training School lasted from 6 to 10 weeks, during which time a trainee studied flight techniques, navigation, geography, math, aircraft recognition, and meteorology. Afterwards, instructors determined whether the trainees would become pilots, navigators, mechanics, or take on other roles.

Potential pilots progressed to Elementary Flying Training School for another 7 or 8 weeks. Here, classroom instruction was combined with at least 50 hours of actual flying time, half of these flying solo. The student pilots learned common flight manoeuvres, taking off and landing, and some aerobatics. For Army Air Service trainees, Elementary Flying training took place at Toronto Island Airport, where the students flew Fairchild airplanes. For the first three classes of Navy Air Service trainees, Elementary Flying Training School occurred at the Toronto Flying Club, with Gipsy Moths, Tiger Moths, and Fleet trainers.

Around May 1941, Initial Training School and Elementary Flying Training School merged into one course. Classroom lessons took place at Little Norway and flying lessons at Toronto Island Airport.

After Elementary Flying Training School, instructors assessed the trainees again. During Little Norway's first few months, successful candidates moved directly on to Advanced Flying School. Army Air Service students attended Advanced Flying School classes at Toronto Island Airport; Navy Air Services students took the classes there and at Patricia Bay, British Columbia.

In April 1941, the training process added a new step: student pilots attended the BCATP's Service Flying Training School after Elementary Flying Training School. Service Flying Training School lasted somewhere between 10 and 16 weeks, including 75 to 100 hours of flying time.

Students selected to fly fighter planes travelled to RAF Station 32 in Moose Jaw, Saskatchewan, and learned to operate single-engine North American Harvards. Students designated for bomber, transport, or coastal patrol duties headed to RAF Station 34 in Medicine Hat, Alberta, and trained on twin-engine Avro Ansons, Cessna Cranes, and Airspeed Oxfords.

Later in the war, students took Service Flying Training School courses at Camp Borden and Hagersville, Ontario.

Successful completion of these advanced courses meant a pilot had earned a two-wing badge, called "wings," and a promotion to the rank of sergeant.

Before heading to combat in Europe, the new pilots joined Operational Training Units, where they gained experience with the aircraft they would be flying in war operations. Some Operational Training Units were located in Canada, such as one at Toronto Island Airport, where pilots flew Curtiss Hawk fighters and Douglas attack bombers. However, most of these Units were attached to RAF air bases in Great Britain.[22]

9

Escape

More and more recruits poured into Toronto's Little Norway. Some were Norwegian patriots who had been living in Canada, or the United States, and had responded to their King's call to arms. But most of the new recruits arrived in Toronto after risking their lives in dangerous escapes from occupied Norway.

Two well-travelled escape routes brought young Norwegians to Canada, one known as the "Easy Way" and the other, the "Hard Way."

The Easy Way

A refugee usually escaped from Norway by crossing its long, forested border into neutral Sweden, a journey taken by great numbers of Norwegians during the summer of 1940. However, Sweden was in a difficult situation: while its sympathies might lie with the Norwegians who wished to flee their homeland, the Swedish feared a German invasion, especially with Hitler's forces in neighbouring Denmark and Norway. To preserve an appearance of neutrality, Swedish police patrolled the border, looking for people illegally entering their country. If found, the escapees were arrested, detained, and, customarily, sent to refugee camps in southern Sweden.

German and Norwegian police forces patrolled the border on Norway's side, where large notice boards warned that anyone caught trying to cross the border would be shot.[1] However, local Norwegian patriots who were familiar with the frontier, known as "border pilots," often guided Norwegian escapees safely into Sweden.

Once in Sweden, Norwegians taking the Easy Way then journeyed halfway around the world, commonly passing through Finland, the Soviet Union, Ukraine, Turkey, Syria, Egypt, South Africa, and South America before making their way to the United States and, finally, Canada, where they joined the Norwegian military forces.

The Easy Way had its lighter moments. The young escapees got to see much of the world. To document their travels, they often collected hotel labels, which they pasted on their luggage. Sometimes they managed to get hotel labels even if they didn't stay at the hotels. A Norwegian airman recalled visiting the Egyptian pyramids, but only after he and his friends had first stopped at a nearby hotel long enough to get their hands on some of its labels.

One young man had no labels on his luggage; when asked why, he said that when he went back to Norway, he'd take only one label on his suitcase—from a hotel in Berlin.[2]

Egil Chramer and Per Hvidsten

Egil Chramer, a dental technician from Fredrikstad, had escaped from Norway with friends. The band crossed through mountainous areas on the way to Sweden, nearly freezing as they slept in the snow, huddling in parkas and sleeping bags. After spending some time in a Swedish refugee camp, Egil set off on his way to Canada, journeying through the Soviet Union, Turkey, and Syria to Suez.

Another Norwegian taking the Easy Way, at about the same time, was Per Hvidsten, a reporter from Sarpsborg. After hiding in the forests around the Swedish border for a period, Per had crossed into Sweden on skis with a group of other refugees, assisted by border pilots.

Both Egil Chramer and Per Hvidsten had served in the Norwegian Army when conscripted some years earlier, though they hadn't known each other. Now, along with some 300 civilians and military personnel, including Resistance fighter Max Manus and about 68 other Norwegian volunteers, they boarded the EMPRESS OF ASIA *in Suez and headed for North America.*

The EMPRESS OF ASIA *was a troopship, owned by Canadian Pacific Steamships Ltd., and had been an ocean liner before being requisitioned for war service. Travelling on the same ship as Egil and Per were 1,000 Italian prisoners of war, kept in the hold. Egil would later recall how he heard the men in the hold singing beautifully every night. The Italian prisoners of war disembarked in Durban, South Africa.*

After two months at sea, the EMPRESS OF ASIA *docked in New York City. Per, Egil, and some fellow Norwegians made their way to Toronto, arriving in September 1941.*[3]

ABOVE

Egil Chramer.

PHOTO: NORWEGIAN ARMED FORCES MUSEUM, OSLO, NORWAY.

RIGHT

Per Hvidsten.

PHOTO: J. PETER HVIDSTEN COLLECTION.

The Hard Way

Escapees travelling the Hard Way sailed in small boats across the treacherous North Sea from Norway to the British Isles. If the sailors were lucky, it would take them two or three days to reach the Shetland Islands; if there were storms and rough seas, it could take 10 to 12 days, or more. A great many Norwegians died during these journeys, killed by mines, enemy airplanes and ships, or simply drowning in the sea.

Crossing the Easy Way. RMS *Empress of Asia* enters New York City's harbour, while Norwegian escapees and other passengers celebrate.

PHOTO: BARBARA CHRAMER COLLECTION.

Oluf Reed-Olsen

One recruit who came to Little Norway the Hard Way was Oluf Reed-Olson, known as "Reed," a 21-year-old from Aker, Norway. Reed and his good friend, Kaare Moe, became saboteurs almost immediately after the German invasion. They began by carrying out small jobs, such as photographing German airfields, for a clandestine British intelligence system in Oslo.

Soon, Reed and Kaare progressed to major missions, including blowing up the Lysaker Bridge, a crucial section of the thoroughfare between Fornebu Airport and Oslo. During a mission to take photos of the inside of a damaged German Heinkel He III airplane parked at Fornebu Airfield, German soldiers arrested them. Reed and Kaare escaped custody by jumping from a moving truck into a river: at that point, they knew they had to leave the country.

The two young men decided to sail to England and then travel to Canada to join Norway's Air Services. As Reed notes in his book, TWO EGGS ON MY PLATE, the first step involved obtaining three crucial things: a crew, equipment, and a boat.[4]

Reed sold his motorcycle to raise money to purchase a fishing smack, an old, leaky, 18-foot-long boat, which the young men christened HAABET, meaning "Hope." Reed was the only one who knew anything about navigation, and that was just what he'd learned as a Boy Scout.

The HAABET cast off from Oslo on September 2, 1940. After a few days of sailing along the coast, the young men were startled to hear an announcement over their wireless radio; the German Command said that any boats trying to escape from Norway would be attacked by every means and sunk without warning.

The next day, a German torpedo boat stopped the HAABET. Reed and Kaare threw out fishing lines and tried to persuade a German lieutenant that they were just fishermen. After sitting down in the HAABET and consuming a good portion of a bottle of whisky, which Reed happened to have brought along, the lieutenant let the boat sail onwards.

While they were stopped at Mandal, a man watched them from the quay. After three days, the man finally approached the boat. He introduced himself as Rolf Gabrielsen, a bank clerk, and asked if they could use a third hand. After checking his identity with friends, Reed and Kaare agreed. The crew was now complete.

They sailed westward for four days. On September 18, the

weather changed. Just 36 hours away from Scotland, the HAABET encountered a violent storm; gale-force winds blew for eight days straight. It was nearly the end for the three men: two of them were swept overboard at least twice and would have drowned if not for their lifebelts—and for safety lines, tied around their waists, securing them to the boat. The boat leaked badly and its pump failed, requiring the men to pump and bail by hand. No one slept. When the sea finally calmed, the men realized that the storm had driven the HAABET southeast, and they were now close to Jutland in Denmark.

Food and fresh drinking water were in short supply, but rather than sail into German-occupied Denmark, they headed back towards England. On September 29, a British patrol plane spotted the little boat, and the men were soon picked up by a destroyer. And not just the men: Reed, Kaare, and Rolf insisted that the HAABET, which had brought them over 1,100 miles of ocean, be saved, too, and the little boat was lifted onto a British ship. Before long, the young Norwegians found themselves in England, where all volunteered to join the forces training in Canada.[5]

Crossing the Hard Way. Oluf Reed-Olsen in the *Haabet*, sailing on the North Sea.

PHOTO: LITTLE NORWAY MEMORIAL.

The Shetland Bus

Fishing boats from Norway were regularly arriving in the United Kingdom by the fall of 1940. The British military began to ask the Norwegian captains of these boats to take part in secret missions. The missions often involved sailing back to their homeland with an undercover agent on board, dropping the passenger off, picking up another one, and then returning to the United Kingdom. The success of these missions resulted in the creation of a base in the Shetland Islands, Scotland, to operate ongoing clandestine voyages to Norway over the course of the war. The route would come to be known as "The Shetland Bus."

These fishing vessels transported agents, ammunition, explosives, radios, and supplies back and forth across the North Sea. The risk was enormous, both from treacherous seas and from attacks by German ships and aircraft. After the Norwegians lost half their fishing boats and crews during the winter of 1942–1943, the American Navy, in fall 1943, provided them with three submarine chasers to replace the fishing boats. These new craft were very speedy, durable, and secure, making operation of the missions much easier.

In addition to secret agents, more than 350 refugees took

the Shetland Bus from occupied Norway to safety in the United Kingdom during the war; many of these refugees made their way to Canada to join the Norwegian air force.[6]

John Stene

In the late fall of 1940, Flight Lieutenant Olav Bakke Stene was visiting New York City. As the United States was still a neutral country, Olav thought he might be able to make a telephone call to Norway from New York. He placed a call to his sister, Helga, and was astonished when it went through immediately. During his conversation with Helga, Olav told her cryptically: "The old family firm is still in business and has openings for more people."

John Stene, Olav's brother, had returned to Norway after delivering one of the Army Air Service's last aircraft to Finland and had resumed work at his old job in Mjøndalen. The Germans hadn't yet taken full control of Norway, but had imposed travel restrictions, blackouts, censorship of radio and newspapers, and rationing of food and goods. Both Helga and John Stene had heard rumours of the Norwegian Air Services re-establishing in Toronto. When Helga relayed Olav's comments to John, he understood Olav's message immediately.

John began planning how to escape to Sweden. With another pilot friend, he took a train to Magnor, the last Norwegian station on the Oslo–Stockholm railway, where his sisters knew a farmer. The farmer told John that the Germans only patrolled the border once each day, at a specific time. By avoiding that time period, the farmer guided the young men safely into Sweden.

John and his friend skied about seven kilometres to Eda, Sweden, and reported to the County Sheriff, who kept them in an unlocked cell overnight. After confirming their identity with the Norwegian Embassy in Stockholm, the County Sheriff released the men. They took the next train to Stockholm. John remembered the excitement of arriving in a fully lit-up city, with stores full of goods, after the blacked-out, bleak streets of Oslo.

As both escapees were trained pilots, the Norwegian Embassy in Stockholm welcomed them. It took about a month and a half for visas to be obtained. During that time, the Embassy housed the men in a small hotel and supplied them with pocket money.

Their journey began with a flight on Swedish Airlines from Stockholm to Moscow, where they spent a couple of days under the care of a Norwegian Embassy official. Next step was a train trip to Odessa, Ukraine, taking two to three days. They left Odessa by boat, travelling to Istanbul, Turkey, where they spent two more days. The ORIENT EXPRESS *carried them to Baghdad, Iraq. After two days there, the men journeyed by an Imperial Airways flying boat to Karachi, Pakistan.*

From Karachi, they travelled by Tata Airlines to Bombay, India, remaining there 12 days while they awaited a ship.

The ship took four to five weeks, with brief stops in South Africa and Trinidad, before arriving in New York City at the end of April 1941. John and his friend took an overnight train from New York to Toronto, and reported for duty at Little Norway.

John later wrote:

> *It was an incredible inspiration to come here and see such efficiency and spirit everywhere. It was also a great joy to meet my brother Olav again and also so many friends and colleagues from Kjeller and Værnes.*[7]

10

Life in Occupied Norway

On April 10, 1940, King Haakon and Norway's Government had told the German Minister to Norway, Curt Bräuer, that they would not accept Quisling as prime minister, as neither the Norwegian people nor the Storting had any confidence in Quisling.

Popular dislike for Quisling's Government had continued in Norway. Public leaders remaining in the country, as well as German officials there, concluded that it would be useful to have a civilian authority established to deal with practical problems arising from the occupation. Norway's Supreme Court negotiated with the Germans respecting this new civilian authority; as part of the arrangement, the Court demanded that Quisling and the Government be removed from office.[1]

The Germans agreed with the Supreme Court's demand and, on April 15, an Administrative Council composed of six leading Norwegians took over day-to-day operation of the country.

Hitler had believed the Administrative Council would work like a substitute government, supplanting the role of the King and Government-in-exile.[2] However, Council members

Vidkun Quisling and German officers during a parade in Oslo, April 20, 1942. First row, from left: Nikolaus von Falkenhorst, Hermann Boehm. Second row, from left: Vidkun Quisling, Josef Terboven, Valentin Feuerstein. PHOTO: NATIONAL ARCHIVES OF NORWAY.

Vidkun Quisling at a Nasjonal Samling youth conference in Borre National Park, Norway, during the summer of 1941.

PHOTO: NATIONAL ARCHIVES OF NORWAY.

insisted on political independence from German control, and a few months later the Council was disbanded.

In the meantime, given Norway's continued war against Germany, Hitler had come to believe that the country needed to be governed with a strong hand. German General Alfred Jodl, Chief of Operations Staff of the Armed Forces High Command, noted in his diary on April 19, 1940: "In the Führer's opinion force has to be used."[3]

On April 24, Hitler named an old Nazi Party colleague, Josef Terboven, to be *Reichskommissar* or Reich Commissioner, and supervise administration of civilian affairs in Norway. Terboven was provincial governor of the Rhineland at the time and had a reputation for ruthlessness.[4] Much to the discomfort of Norwegians, he moved into *Skaugum*, the Oslo residence formerly inhabited by Norway's Crown Prince Olav and his family.

General Nikolaus von Falkenhorst remained in command of the German Army stationed in Norway, but Terboven controlled the secret police.

On September 25, 1940, the Administrative Council was replaced with a Council of Commissioner Ministers to govern Norway in consultation with Terboven's own administrative group, the *Reichskommissariat*. Ten of the 13 Commissioners were from the Nasjonal Samling Party, but Quisling was not included. At the same time, Terboven abolished all political parties except the Nasjonal Samling, and it grew in membership.

About a month later, the Minister of Justice tried to infringe on the independence of the courts by giving himself sole power to appoint jury members, assessors, and people holding other judicial offices. After denouncing this action to no avail, the Supreme Court justices all resigned.

Resistance begins

Even before the war in Norway had ended, the British had begun to recruit Norwegian agents, like Oluf Reed-Olsen, for secret operations against the Germans. In July 1940, the British established an organization called the Special Operations Executive (SOE), to organize and assist Resistance movements in occupied countries: by the fall, a handful of Norwegians were being trained as SOE agents. Lieutenant Martin Linge, a former actor, became the leader of the Norwegians in the SOE; his group was known as the Norwegian Independent Company No. 1 or "Company Linge."[5]

At the same time, the Shetland Bus operations had begun, bringing agents across the North Sea to and from Norway.

While Norwegians themselves remained fairly passive during the summer and fall of 1940, small acts of resistance occurred. Former soldiers hid their weapons where they could find them again and kept in touch with each other. People threw rocks at German soldiers and tore down Nazi posters. Some wore safety pins or carried paper clips symbolizing Norwegians' unity, in the sense of being "clipped together."

The German occupiers censored newspapers and radio broadcasts, but underground leaflets materialized, spreading patriotic messages.

King Haakon and the Government-in-exile kept in contact with Norwegians through radio broadcasts from London via the British Broadcasting Corporation (BBC). One such example was a letter the King read over the BBC on July 8, 1940. His letter was a reply to correspondence he'd received the previous month from some members of the Storting who had remained in Norway. Under duress, they had asked him to abdicate. In his broadcast, the King pointed out that the Storting had written to him under threat and intimidation, rather than as a result of their free deliberations. King Haakon rejected the idea of abdicating, saying it was a betrayal of his constitutional duty to his people. His words found great welcome in Norway, as they quickly and secretly spread through the occupied country.[6]

Norwegian women protest Nazi occupation. In English, the words written on the back of the woman on the left mean "down with the N.S." (Nasjonal Samling.) The H7 symbol on second woman's back is a tribute to King Haakon VII. PHOTO: NATIONAL ARCHIVES OF NORWAY.

Rolf Hauge

Rolf Hauge had been imprisoned at Akershus Fortress since being arrested by the Germans in his Finnish Army uniform. Once Norway surrendered, however, he was freed along with other political prisoners. Before the Germans released them, the prisoners had to promise they would not fight against the occupiers again. Like most of the Norwegians being freed, Rolf signed the German document, but never intended to abide by its terms.

Instead, he worked with the Resistance, cycling around the countryside and furtively retrieving weapons hidden by Norwegian military troops just before they were demobilized. The Resistance told the young men where to look for the weapons. When the guns were found, Rolf and his friends hid them in their backpacks and cycled along, trying to look like tourists.

Trouble developed after Rolf purchased a rifle from a Norwegian man, which he then delivered to the Resistance. Ownership of a firearm was restricted by law. When the gun seller was subsequently arrested by the German authorities, he divulged, under interrogation, that he'd sold a rifle to Rolf. The Gestapo came to Rolf's home looking for him but Rolf, like many young members of the Resistance, was living in the forest. The Gestapo told Rolf's father that if his son was not at home the next time they called in, the father himself would be arrested.

It was lucky that Rolf was not at home when the Gestapo visited, as he had a rifle hidden in his backpack.

To protect his father, Rolf made sure he was at home the next time the Gestapo came by, and he was arrested and taken to prison. It was important to pretend to cooperate with the Gestapo, and Rolf made up a story about having already sold the rifle to someone he thought to be a sailor, though he didn't know the sailor's name or how to locate him.

While incarcerated, the prisoners worked under the watchful eyes of armed guards. One of Rolf's tasks was to collect books and other material that the Germans had confiscated from the populace as being unsuitable. He took the books to a truck, which conveyed them to a paper factory in Drammen, where the material was destroyed.

One day, by a stroke of good fortune, Rolf recognized the driver of the truck as someone he knew. Rolf smuggled a note to the truck driver, asking him to go to the Gestapo and report that the fellow who sold Rolf the rifle had just heard of Rolf's arrest, and had some information for the police. The seller then wrote a note to the Gestapo, confirming that Rolf had sold the rifle to a third party and providing an address where they could find the gun. In the meantime, the Resistance placed another gun at that address, and when the Gestapo found the weapon, they believed that Rolf had been telling the truth all the time, and he was freed.

But even when released, Rolf had to report to the Gestapo regularly and was kept under surveillance. He found work as a night watchman for a company that provided security services to businesses and industry. Rolf's new employment was lucky for the Resistance, as it gave him knowledge of many places that might be good targets for bombing and sabotage; knowledge that he passed along to the Underground.

It was getting dangerous for Rolf to remain in Norway. Many people working with the Resistance were being sent to German concentration camps, and he felt his own luck was running out. He decided to head for Sweden, along with three friends.

They planned to travel to Sweden by train, bus, and on skis. First, they needed to find enough food for the three- to four-day ski trip. A Nazi-sympathizing baker, living in Drammen, owned a big cottage in the country and the men thought the baker must have had lots of food stored there. They broke into the basement of the cottage while no one was home and helped themselves to enough canned food for the journey to Sweden.

The first leg of their journey, by train, was uneventful. The second leg was by bus. At the end of the route, the Gestapo stopped the vehicle. The soldiers asked the men where they were heading. It was Easter, a time when many Norwegians travelled to mountain locations to ski, so Rolf and his friends said they were going skiing and showed the Germans their travel passes. After a cursory glance, the Germans let them continue: this was fortunate as the travel passes didn't actually include the territory in which they were travelling.

The young men walked down the road until dark, then headed into the forest. They skied towards Sweden, staying away from well-

travelled roads. Rolf had a compass and was familiar with the area, but still the going was difficult as the snow was heavy and wet.

It took four days to reach and cross the Swedish border. By then, the snow had disappeared so Rolf and his friends abandoned their skis and walked on the roadway. They were hoping to find a bus or train station where they could buy tickets onward, but the Swedish border patrol had seen their tracks. They were picked up and taken to southern Sweden, to a big camp that housed many other Norwegian refugees. The camp was well organized, and the refugees were allowed to work in the Swedish forests, which Rolf and his friends did for a while. However, as they wanted to travel to England, join the Norwegian Armed Forces, and get back into the war, the group soon contacted the Norwegian Embassy in Stockholm and journeyed to that city.

While in Stockholm, Rolf signed up to take courses at a technical school but, in fact, spent most of his time working for the Resistance. He was paid five kroner a week, and he recalled that it was just enough to buy a beer in a restaurant.

As Rolf had already been trained by Norway's Navy Air Service, Norwegian authorities soon flew him to Leuchars, Scotland; from there, Rolf continued his journey to Canada.[7]

11

Flying High

Each morning, Little Norway personnel took the ferry across the Western Channel to Toronto Island Airport. In winter, this could be a trying experience: Harald Jensen described the passengers' hands being chafed and their noses turning blue from the raw March winds as they sailed along.[1]

At the airport, mechanics examined the planes to be used that day before the instructors and student pilots climbed into the aircraft.

Curtiss and Douglas aircraft lined up at Little Norway, Toronto Island Airport.

PHOTO: NATIONAL ARCHIVES OF NORWAY.

The first serious accident occurred October 23, 1940, when an aircraft bearing two Norwegians crashed into a corn field near Edgeley, Ontario. One of the passengers was instructor Per Waaler, who had shot down a German plane at Fornebu Air Base on April 9, 1940. Waaler had escaped from Norway by ship in May 1940, and then travelled to Little Norway, becoming a flight instructor.

Waaler and student pilot, Arnsinn Brodersen, left the Toronto Flying Club in a Gipsy Moth and were practising emergency landings. A neighbour who witnessed the accident said, "The plane was flying low. Its nose just seemed to turn down and fall."[2]

A number of witnesses raced to the crash site where they found the plane bent in two with its tail sticking up in the air. The neighbours pulled Waaler from the rear seat and then discovered Brodersen in the front seat. "We had to rip everything apart to get him out," reported one witness. Brodersen suffered severe chest injuries, several broken ribs, and a fractured leg. He was unconscious for over eight days. Waaler broke his leg and had head injuries. Luckily, both recovered.[3]

Weather could cause difficulties for the new pilots. Flight Lieutenant Kaj Birksted was piloting a training plane on December 26, 1940, in a dense fog. He was taking Sergeant John Lome to St. Thomas to make funeral arrangements for his late wife, who had passed away a couple of weeks after having given birth to their daughter. The poor weather conditions forced Flight Lieutenant Birksted to try landing the aircraft in a Hamilton park. As some residents watched anxiously, the plane circled an open stretch of Eastwood Park and dropped slowly. Birksted landed the aircraft, but the plane charged towards a bank of trees, crashing into one.

Lome was trapped in the wrecked fuselage. "I thought it was good-bye for a second or two after it crashed," Lome told a reporter. "My first thought was that I was properly trapped if the plane broke into flames. That fear passed when I saw Lieutenant Birksted jump out like a flash. Then some men started to free my legs by cutting away the bars with a hacksaw," he said. Both flyers were treated in hospital where Lome was found to have sprained both his ankles. Nevertheless, he continued on by train to St. Thomas to make the funeral arrangements.[4]

More planes were arriving at Toronto Island Airport: by the end of 1940, Little Norway had 12 Fairchild PT-19 trainers, 32 Douglas 8A-5 attack bombers, 6 Curtiss Hawk pursuit planes, 4 Stinson Reliant seaplanes, and one Waco E-Series biplane. Not all had been assembled, however.

Around February 1941, additional aircraft were delivered. These included four Douglas 8A-5s, 30 Curtiss Hawks, and one more Stinson Reliant.[5] In fact, there was so much aircraft activity at the Toronto Island Airport that three zones were set up: taxiing, take-off, and landing. A pilot couldn't take off or land without approval.[6]

Each day, Toronto residents heard the aircraft roaring above them. The noise began around 9:30 a.m. and continued all day, with pilots seeming to take time out from flying only for a lunch break.[7]

Cyril Fry

Gravenhurst's Cyril Fry, then a teenager living in Toronto, had a keen interest in aircraft. Although he had never flown, he often rode his bike to local airports, where he enjoyed watching the planes. Fences were rare, and he was usually the only spectator.

After the RNAF arrived at the Toronto Island Airport, young Cyril and his bike could sometimes be seen at the waterfront, looking across the Western Channel at Norwegian aircraft on "circuits." These are the repetitive take-off, circle, and landing exercises, which teach some of the most challenging parts of flying. He remembered seeing the yellow and blue Fairchilds flying above, but the biggest thrill was watching the RNAF's big floatplanes taxi out in the harbour, establish wind direction, accelerate and lift off in a cloud of spray, circle, check the wind direction again, throttle back and, hopefully, gently settle on the surface. As Cyril noted, manoeuvring aircraft on water is a different challenge than on land, because the effects of wind and waves are more complex, and steering more difficult.[8]

Don't hit the buildings

One reporter who rode in a Douglas bomber, which was part of a group of bombers practising manoeuvres, wrote that when the Norwegians referred to flying in "tight formation," they meant what they said. "An agile man could have walked from wing tip to wing tip."[9]

Some of the pilots with pre-war experience, and now flying modern aircraft for the first time, found it thrilling to fly low. Lieutenant Harald Olafson, who had been named Chief of Instrument Flying at Little Norway, was concerned about

Royal Norwegian Navy Air Service pilots training at Toronto Flying Club, December 1940. Oluf Reed-Olsen is fourth from left. He wrote the names of other pilots on the photo, drawing a cross beside the names of those subsequently killed during the war. PHOTO: OLUF REED-OLSEN FAMILY COLLECTION the young men's overconfidence. Not altogether jokingly, he admonished the flyers to avoid hitting Toronto's buildings.[10]

Little Norway was becoming known worldwide, and soon Norwegians from many places became interested in joining the Air Services training in Canada. The Government-in-exile also imposed conscription in an attempt to ensure a steady flow of recruits: in June 1940, all Norwegians living in the United Kingdom, between the ages of 21 and 35, were called up for military service. The next year, conscription was extended to certain other designated countries, including Canada, Cuba, Brazil, Argentina, Mexico, Australia, South Africa, India, China, Egypt, Portugal, Iceland, and the Faroe Islands.[11]

The Government-in-exile couldn't enforce conscription on Norwegians living in all foreign countries and largely relied on the patriotism of such young Norwegians in answering the call to military service.

Odd Toldnes

Odd Toldnes was born in 1919 in Porsgrunn, Norway, but his family moved to Brazil around 1930. Odd grew up in Sao Paulo. Responding to the Government-in-exile's conscription summons, Odd decided to join the RNAF. *To get to Canada, he signed on as a cabin boy on a Norwegian merchant marine ship, the* TRAFALGAR, *which sailed from Buenos Aires to New York City and arrived there in February 1942. Odd's timing was fortunate, as a few months later, a U-boat sank the* TRAFALGAR *near Guadeloupe.*

From New York City, Odd travelled to Little Norway in Toronto, where he became a member of Little Norway's ground crew as a mechanic and aircraft painter.[12]

Odd Toldnes.

PHOTO: KAREN GREAVETTE COLLECTION.

12

Toronto Welcomes the Norwegians

Ever since April 9, 1940, Norwegians and other Scandinavians residing in Canada had wanted to help the refugees, who had fled Norway and were living in England. To this end, a group, including Carl Printz, Norway's Vice Consul in Toronto, had established a fundraising organization called the Norwegian War Aid Fund Association of Ontario. The Fund's Ladies' Auxiliary contributed clothing for the refugees through its sewing and knitting circles.

Newspaper articles regularly published news of dances, bazaars, concerts, and other social events held by the Fund. The press generally sympathized with Little Norway, printing articles about good-looking young airmen recounting stories of their dangerous escapes from a Nazi-occupied homeland.

The Norwegians' bravery also endeared them to Torontonians. When a 70-mile-an-hour storm ripped through Toronto on September 25, 1941, Norwegian trainees rushed to the aid of yachtsmen struggling to save their boats. Some swam through the turbulent waves with tow ropes to help haul the boats to shore. "We have a storm like this every two weeks on the west coast of Norway where I come from," one declared to a reporter.[1]

Before long, social invitations to dinners at private homes, dances, and parties began to arrive at Little Norway. Dagmar Hinch, daughter of Carl Printz, was one of the first to welcome groups of Norwegian officers to her home for parties; she also arranged parties for them elsewhere. On October 3, 1940, Mrs. Hinch told a *Toronto Daily Star* reporter that the gatherings were informal, where only first names were used because the last names were so difficult. She added that she wasn't really needed anymore: the young men had come to know so many Torontonians that they went partying by themselves.[2]

One flight instructor recalled, "On Monday morning there would be a little boasting between the officers about how many cocktail parties they had managed on Saturday night."[3]

Many airmen stationed at Little Norway dated Canadian women, and it didn't take long before marriage announcements began to appear in local Toronto newspapers. The first marriage reported in the newspapers took place on December 26, 1940, between Mary Loosemore of Toronto and Erling Jørgensen of Oslo. Jørgensen, formerly a medical student in Norway, had fought the Germans, been captured and imprisoned, then escaped from a Nazi prison camp. With a friend, he had walked 177 kilometres across Norway and Sweden to reach Finland, and from there, made his way to Canada, joining the Navy Air Service.[4]

When another pair married soon after, a reporter wrote: "Another whirlwind romance between a Norwegian flier and a Toronto girl will lead to the marriage on Jan. 8 of 18-year-old Jeanette Stephens ... and Aage Schaanning at St. John's Garrison Church. It will be the second marriage between a Toronto girl and a Norwegian flier within two weeks."[5]

And the marriages kept coming. Lieutenant Harald Olafson married Anita Falla of Toronto on February 18, 1942, at St. John's Garrison Church. As the newlyweds left the church, other Norwegian airmen formed a guard of honour.[6]

Sergeant Arne Skretteberg also met his future bride in Toronto. A friend of Skretteberg's was dating a girl who worked in a Honeydew restaurant in the city, and one day he brought Skretteberg along to the restaurant for breakfast. There, Skretteberg met a lovely young waitress, Irene Czajkowski. They began to go out together, often on double-dates with the other couple. Less than a year after meeting, they married.[7]

The overwhelming popularity of the Norwegians with Toronto's young female population led to occasional conflicts with Canadian men. Fights sometimes broke out in spots like the beer parlours of the Piccadilly Hotel on King Street West, a favourite spot of the former whalers.[8]

One RNAF airman's son recalled speaking with a Canadian man, who had been at a bar in a downtown Toronto hotel with six friends in 1942. The group thought that some women were paying too much attention to a couple of Norwegian flyers and concluded that the best way to get rid of the Norwegians was to fight with them. In a subsequent confrontation, the seven men beat up the two flyers. Luckily, such one-sided fights seem to have been rare.

Most of the Norwegians were keen to avoid trouble in case their superior officers heard about it and decided they weren't fit to progress in their training. On the whole, discipline wasn't a serious problem at the camp.

Sometimes the recruits slipped up. During June 1941, the Toronto Harbour Commissioner's night watchman reported that on many occasions, the Norwegian sentry had abandoned his position and gone inside a building. Once, the sentry was found asleep in a truck, having left his rifle standing against a wall outside, some six to nine metres away from where he lay napping.[9] On another occasion, the night watchman couldn't find anyone guarding the Curtiss and Douglas aircraft at the Toronto Island Airport. He went looking for the sentry and finally found him, lying asleep at the doorway to one of the buildings.[10]

Douglas attack bomber.

PHOTO: LITTLE NORWAY MEMORIAL.

There were also reports of the "promiscuous" shooting by sentries, endangering the lives of the lighthouse keeper and his assistant, who regularly had to cross the airport grounds.[11]

Little Norway's commanders soon realized that the recruits did not have sufficient experience to properly serve as sentries or deal with lawless behaviour. They decided to create a trained security service.

By March 1941, a service police force had been set up, composed of some 56 individuals who had taken policing courses, including military law, use of firearms, boxing, judo, wrestling, and investigations. The camp also acquired a police cruiser with radio equipment.

Little Norway's senior officer had the power to award punishment for breaches of discipline and various misdemeanours. When an offender received a sentence of less than 15 days, the individual served the time in one of six detention cells at Little Norway. For sentences of over 15 days, an offender was placed in detention cells located on the *Iris*. After the *Iris* sailed away on August 22, 1941, the Norwegians sent those with the longer sentences to the Canadian Military Detention Barracks.[12]

Johan Egon Malmsten,
a member of Little Norway's Service Police Force.

PHOTO: ELAINE HOOPER COLLECTION.

A soldier for Christmas

Toronto's private citizens and charitable organizations didn't forget the Norwegian airmen at Christmas time.

Rolf Kolling recalled being one of about five men from Little Norway who attended a party at the home of Sir Ernest MacMillan, a famous orchestral composer, conductor, and musician. Mrs. MacMillan made them feel at home; along with some 20 other guests, the young Norwegians enjoyed dinner, played records, and danced. There was also a big barrel of beer. Kolling, who didn't drink alcohol, had to remind the other four student pilots to go easy on the beer, as it was meant to serve 25.[13]

The West Toronto Kiwanis Club Ladies Auxiliary held a supper for Canadian and Norwegian airmen at the West End YMCA. Fifty from Little Norway came to the supper in 1940. The group sang carols and Norwegian folk songs, and every diner received a Christmas parcel. On December 23, Howard Park United Church entertained 250 Canadian and Norwegian airmen, who watched as 15 turkeys, each weighing about 10 kilograms, were carried into the room to the music of bagpipes.[14]

The YMCA ran an annual campaign to ensure that every military man away from home, but living in Toronto, received an invitation to Christmas dinner at a private citizen's residence. This "Soldier for Christmas" initiative encouraged householders to invite two military men for December 25th dinner. These campaigns were wildly successful: in 1940, 771 men were treated to dinner in 348 homes. In fact, there hadn't been enough soldiers to go around as some 850 requests couldn't be fulfilled. At Little Norway, the excess invitations were held over for New Year's Day.[15]

It's a Norwegian tradition to celebrate Yule on Christmas Eve, rather than on Christmas Day, and that was no different in 1940, even though the airmen were so far from home. Christmas trees were decorated with artificial white snow and white ornaments, vividly contrasting with the deep green branches of the tree. Captain Riiser-Larsen explained that the colour white signified Norway's innocence of guilt in this world war and the colour green represented Norwegians' hope of returning to a peaceful and free Norway.[16]

Following dinner on Christmas Eve, the Norwegians danced around the Yule tree in the traditional manner, and sang Christmas carols. Messages from the King and Crown Prince were read out to the assembly.

Mrs. Riiser-Larsen had taken on the task of ensuring that every person at Little Norway received a Christmas gift. Two months before Christmas, she started preparing the packages in her apartment but soon ran out of room. Canadians and Americans of Norwegian descent contributed so many gifts that space had to be rented elsewhere to store the packages. In the end, each person received several gifts.

In 1941, the men got so many gifts from Norwegian friends in Canada and the United States that they sent three-quarters of them to sailors in the Royal Norwegian Navy and merchant marine.[17]

On Christmas Day, many of the servicemen dressed in their best uniforms and left to enjoy dinner at private homes, including those of Torontonians who had taken part in the YMCA's Soldier for Christmas initiative. The men were grateful for these invitations; as one trainee remarked, "a mess hall, however friendly, is no substitute for a home."[18]

At New Year's in 1940, one-third of the airmen travelled to Montreal, on furlough, for a ski holiday at the Government-in-exile's expense. Everyone remaining at Little Norway enjoyed a full festive dinner on New Year's Eve, followed by dancing.

Family

It must have been very worrisome for the Norwegian recruits to think about their families living in a homeland occupied by the enemy; it must also have been very worrisome for those families to think about their absent children, living somewhere unknown.

Families back in Norway usually had no idea exactly what their fugitive children were doing or where they were stationed as it was too risky for RNAF personnel to write home. If the Gestapo found out the identities of the escaped flyers' parents, brothers, sisters, or other family members, those individuals would be in great danger.

Sometimes recruits who had recently arrived from Norway had information for RNAF personnel about their family members back home. Sometimes those family members managed to send letters to the trainees, usually through the efforts of the Red Cross. Rolf Kolling recalls receiving five or six letters from home over the course of the war. The messages were kept very short, such as "We are all well," in case the letters fell into enemy hands.[19]

Frances Malmsten, Egon Malmsten, and Isabel Dinsmore.

PHOTO: ELAINE HOOPER COLLECTION.

Isabel Dinsmore and Clementina Whiteside

Isabel Dinsmore lived on Lakeshore Avenue, Centre Island, and often saw the young recruits marching by her home. Perhaps because her own two sons were serving in the Canadian Navy, Isabel invited some of the Norwegians to her home from time to time. She played the piano and could do so by ear; as the young men sang songs from their homeland, she would pick up the tunes and accompany them. The recruits enjoyed the family atmosphere and Isabel sometimes felt like a surrogate mother. In time, Isabel's daughter, Frances, began to date one of the recruits, Johan Egon Malmsten. Egon was Swedish and had sailed with Norway's merchant marine before the German invasion. At Little Norway, Egon worked as a member of the Service Police Force.

In 1943, Egon and Frances married at the Royal York Hotel.[20]

Another Torontonian who welcomed the young Norwegians into her home was Clementina Whiteside. Clementina lived in The Beaches area of Toronto and was raising three children: Joan, who was around 18 years old; Bill, who was 15; and Ross, 13. She heard about the loneliness experienced by some of the young men at Little Norway and called the camp's headquarters, volunteering to have two of the recruits come to her home for Sunday dinner. This was especially generous of her, as her husband had died suddenly in 1936 and money was tight. Mrs. Whiteside was already taking in boarders to help meet expenses.

After an enjoyable evening with the first two Norwegians, Mrs. Whiteside invited two more young airmen to dinner, then two more, and so on as the war continued.

Bill Whiteside recalled the Norwegian visitors as being very enjoyable company. They dressed in uniform, spoke English well, told stories of their homeland's beauty, laughed a lot, and, in short, made a great impression. Although the conversations didn't particularly focus on the war, it was clear to Bill that the young men were eager to get back overseas and fight the Germans.

One of the early visitors to the Whiteside home told the family about an airman who had been very involved in Boy Scouts in Norway and who wanted to get in touch with Canadians who shared his interest. As both Bill and his brother Ross had been Boy Scouts, Mrs. Whiteside invited the airman to dinner.

It was Oluf Reed-Olsen who arrived at the Whiteside home one Sunday evening. Reed was handsome and charismatic; over dinner, he told the family how he and his friends had escaped from Norway in a fishing boat. The Whitesides enjoyed Reed's company, no one more so than Joan, and he was invited back for dinner many times. Reed and Joan began to date.[21]

Clementina Whiteside.

PHOTO: SUSAN WHITESIDE DAVIS COLLECTION.

NORWAY

13

Entertainment

Most of the former whalers and sailors had lots of money to spend in Toronto, as whaling was a lucrative trade and their salaries had accumulated from their last, pre-war work at sea. These men often bought themselves cars and enjoyed a lively social life in Toronto.

But many air force recruits had arrived in Toronto virtually penniless. A recruit was paid $1.00 per day, a salary that didn't stretch very far.

The Royal York Hotel in Toronto was a very popular place to spend off-duty hours. The airmen had to pay the hotel's cover charge to hear the bands, but they sometimes saved on costs by sneaking in their own bottles of liquor. An airman would order a glass of pop, or other mix, and add alcohol to the glass from the bottle hidden under the table.[1]

Sometimes, a recruit had just enough money to take the bus from Little Norway to the Royal York Hotel. That didn't mean the night would be dull, however. Parties were always taking place at the hotel, and once the partiers saw a trainee's Norwegian uniform, someone would usually call out: "Hey Norway, come and have a drink."

The Active Service Canteens welcomed troops from all the armed forces with entertainment, food, and hospitality. One Active Service Canteen, located at 12 Adelaide Street East, Toronto, was staffed by some 600 female volunteers per week, working different shifts in a day that began before 8 a.m. and ran until midnight. They served as waitresses, administrators, cashiers, cloakroom attendants, hostesses, and kitchen staff. Most of the women who worked in the evenings had full-time jobs during the day.

Cooks and volunteers at the Adelaide Street East canteen turned out hearty, inexpensive meals for thousands of military personnel each month. A diner could enjoy a meal such as roast lamb with mint jelly, vegetables, bread and butter, a beverage, and pie, all for the price of 25 cents.[2]

Families and girlfriends of armed forces personnel could join them for a meal or entertainment at the canteen. On Sundays, when touring theatre groups or local performers put on free shows, overflow crowds often had to be turned away.

Wednesday and Saturday nights were for dancing. No one was left out: volunteer companions, identified by flowers

OPPOSITE Royal visit, January 1941.
From left: Lieutenant Colonel Ole Reistad, Crown Princess Märtha, Commander Fredrik Vogt, and Crown Prince Olav, speaking with an airman.

PHOTO: ANONBY FAMILY COLLECTION.

pinned to their shoulders, acted as partners for any servicemen who arrived alone. Two young volunteers, who estimated that they'd danced 1,000 miles with servicemen since war began, told the *Toronto Daily Star* in 1944 that sailors were the best dancers of all, no matter where they came from. But when it came to waltzing, one of the young women declared that Norwegians were especially good—although she admitted her objectivity might be affected by the fact she was engaged to marry a Norwegian.[3]

Tuesdays and Thursdays were for card games such as bridge. Some enjoyed the canteen's billiard and ping-pong tables, while others found quiet places in the lounge to write letters.

However, there were rules against members of the armed forces dating the canteen volunteers, rules that sometimes had to be worked around by arranging meetings on the sly.[4]

The YMCA made its sports facilities freely available to visiting airmen from all nations. Toronto's Central, West End, and Broadview YMCA branches held Sunday School gatherings and, afterwards, provided a meal where some 50 to 60 girls acted as supper companions.

In addition, the Norwegians were generously entertained by local citizens, such as the Toronto family who drove Rolf Kolling to visit Niagara Falls one day.[5]

TOP Airmen beside Plymouth. Jakob Anonby wrote the men's names on the back of the photo. From the left: "W. Nilsen, Holthe, Waagen and Anonby." PHOTO: ANONBY FAMILY COLLECTION.

BOTTOM Active Service Canteen at 12 Adelaide Street East, Toronto, on October 30, 1943. From left, an Active Service Canteen customer, Convenor Molly Parker, and volunteer, Kathleen Hinton. PHOTO: CITY OF TORONTO ARCHIVES, FONDS 2, SERIES 958, FILE 52.

The Norwegians entertained themselves, too. A soldiers' club, "The Flying Norseman," created early in 1941, located right at the camp, provided a gathering place where the men held sing-songs, played games, and listened to lectures.

Dances were popular on Saturday nights at Little Norway. The former Laura Merritt remembered going to dances and movies with her Norwegian boyfriend, Jakob Anonby, as well as spending evenings with him at her parents' home. Jakob sailed with the merchant marine before the war and, at Little Norway, was learning flight mechanics. After he was posted overseas, Jakob wrote regularly to Laura.[6]

Choir

Early in 1942, Second Lieutenant Finn Andvig formed the Norwegian Male Choir at Little Norway. Up to 40 men joined initially, though their number fluctuated as members left for overseas and new replacements had to be found. The choir met regularly on winter evenings to sing favourite Norwegian folk and traditional songs. Toronto choral groups like the Lyric Male Choir invited the Norwegian Male Choir to join them in concerts on various occasions.[7]

Crown Princess Märtha and Crown Prince Olav visit shoe repair shop during tour of Little Norway, January 1941. PHOTO: ANONBY FAMILY COLLECTION.

Royal visitors

The first visit by Crown Prince Olav and Crown Princess Märtha to Little Norway took place on January 13 and 14, 1941. The north wind was bitingly cold on January 13, as the camp prepared to greet the Royal couple.

Olav wore a grey Norwegian Army uniform; Märtha kept warm in a mink fur coat and woollen stockings. They entered and inspected every building at the camp, asking questions and speaking to all they met. Most of the Norwegians at Little Norway had never been this close to the Crown Prince before and were very excited. For some, the experience was overwhelming. One cook became so overcome with the honour that he withdrew to his room and wouldn't return.

The Crown Prince addressed a joint luncheon meeting of the Canadian and Empire Clubs, then the Royal pair travelled to Toronto Island Airport to watch the Norwegian student pilots perform aerobatics. Curtiss Hawks and Fairchilds delighted the viewers, especially when a formation of 12 blue and yellow Fairchilds dipped their wings in salute.[8]

The Air Services trainees marched to St. John's Garrison Church, located at the corner of Wellington and Portland Streets, for a special service attended by the Crown Prince and Crown Princess. Although most of the Norwegians were Lutherans, this Anglican Church served the military regiments.

Following the church service, Olav and Märtha spent the evening dining with men from all ranks in the big mess hall at Little Norway. The Royal Family made many visits to Little Norway over the years, and their relationship with the flyers was always a friendly, informal one.

On January 14, another bitterly cold day, the Royal party visited the Toronto Flying Club and inspected the third class of naval aviators to graduate from the Club's Elementary Flying Training School. Pilots in nine open Tiger Moth airplanes gave a brief flying demonstration. In the evening, Lieutenant Governor Albert Matthews, and Mrs. Matthews, held a dinner party in their Queen's Park suite in honour of the Crown Prince and Crown Princess.[9]

As well as Norwegian royalty, a steady stream of visitors came to Little Norway, including members of the Government-in-exile and politicians. Wendell Willkie, who assisted U.S. President Franklin D. Roosevelt as an informal ambassador, visited Little Norway in 1941, when in Toronto to support Canada's War Services Fund Drive.[10] The Danish Ambassador to the United States, Henrik Kauffmann, came to the camp to inspect and meet the Danes who had joined the Norwegian Air Services. Other visitors included Prince George, the Duke of Kent, and Florence Jaffray Harriman, American Ambassador to Norway.

14

First Fatalities in Canada

The extremely cold winter of 1941 meant that ice on Toronto harbour was too thick for floatplanes to take off and land, and so the Advanced Flying School for Navy Air Service students took place in Patricia Bay, British Columbia. Six Northrop patrol bombers and a Stinson Reliant airplane were shipped there for the student pilots' use.[1]

Airman Erling Jørgensen, who on December 26, 1940, was the first Norwegian trainee to marry a Toronto girl, was, sadly, also one of the first to die on Canadian soil. Jørgensen and Sub-lieutenant Harald Kruse had been practising landings at Burrard Inlet, Vancouver, on February 21, 1941. According to witnesses, their Northrop seaplane came in at too sharp an angle and sank immediately.[2]

A second fatal crash occurred in Patricia Bay on March 18, 1941, when Lieutenant Kaare S. Kjos and Sub-lieutenant Jens E. Riiser were killed. Their Northrop seaplane apparently slipped out of a low-altitude turn after take-off and plunged into the sea. A third trainee, Lieutenant Erik Bjørneby, suffered a broken back but survived.[3]

When the trained Navy Air Service pilots returned to Little Norway, the remaining four Northrops accompanied them back to Toronto Island Airport.[4]

Toronto Island Airport had its first fatality on March 26, 1941, when a 23-year-old Danish student pilot, Corporal Preben Malm Rasmussen, died after his Fairchild aircraft dove into Lake Ontario. It was reported that Rasmussen had been practising aerobatics before the crash and may have been about to land when his plane went into a spin from an altitude of about 300 metres. "It was spinning like a leaf in a high wind," said a witness. "I thought it was stunting, but it just kept on coming down."[5]

The accidental deaths of young Norwegian flyers were never easy to accept. The Air Services held an inquiry after each accident in an attempt to determine the cause and endeavour to prevent a similar occurrence in the future.

Funerals were customarily held at St. John's Garrison Church. Within days of an airman's death in Canada, the information travelled to the United Kingdom, then Sweden, and finally to his family in Norway, who usually heard the news from their local minister. Unless the airman was buried in Canada, his cremated remains were returned to Norway after the war.[6]

British Commonwealth Air Training Plan courses begin

On February 17, 1941, the first Norwegian Air Services recruits attended RCAF courses, given by the BCATP. Four student navigators and two officers travelled to Mossbank, Saskatchewan, and to Jarvis, Ontario, to study at Bombing and Gunnery Schools.[7]

Airmen in cockpit, Little Norway, November 1940.

PHOTO: ARCHIVES OF ONTARIO, C 109-2-0-18.

John Stene had been appointed Senior Flying Instructor as soon as he arrived at Little Norway in May 1941, replacing Wilhelm Mohr who had been posted overseas. Stene and another officer brought some 17 graduates of the Elementary Flying Training School to Moose Jaw, Saskatchewan, for Service Flying Training School. The Norwegians' training on big Harvards went smoothly and the airmen did well.[8] In fact, the Norwegians always did very well at BCATP courses, customarily graduating at the head of their classes.[9]

Squadrons heading overseas

The Air Services' chief goal had been to create Norwegian air squadrons and send them into battle as soon as possible. By late winter 1941, the *Evening Telegram* noted Torontonians had been wondering when the first squadron would leave for overseas. The writer pointed to the great strides that had been made in the Norwegians' flying skills:

> Their formation work over Toronto recently has been something to gape at, and their aerobatics with the Douglas dive bombers and the fast Curtiss P-36 have been a joy to watch. They will give good accounts of themselves when they reach England to help the RAF.[10]

April 25, 1941, marked a crucial stage in Little Norway's evolution; on that day, the first squadron left for active duty overseas. This unit of some 29 naval airmen and officers, who were making their way to Iceland, included pilots who had graduated from the BCATP course at Patricia Bay, and had continued with Operational Training at Toronto Island Airport during the spring. The squadron also included naval navigators, air observers, air gunners, mechanics, and other ground staff. With the addition of some Norwegian naval pilots and sailors already in the United Kingdom, this group formed the 330 Norwegian Squadron of the RAF.[11]

Little Norway sent three Northrop twin-float seaplanes to Iceland for this squadron to use, along with equipment and spare parts for the aircraft. The squadron operated under the RAF's Coastal Command from three Iceland sites, with headquarters at Reykjavik, and began carrying out reconnaissance and convoy escort duties in the Arctic. The squadron would move to Oban, Scotland, in 1943.

In June 1941, a squadron of Army Air Service pilots, air and ground crew, travelled from Little Norway to Catterick, England. They joined the RAF's Norwegian fighter plane unit, the 331 Norwegian Squadron, established on July 21, 1941.

From then on, air crew regularly left for overseas duties.

A Northrop N-3PB of the Norwegian No. 330 Squadron,
Royal Air Force Coastal Command, in Iceland, October 1941.

PHOTO: IMPERIAL WAR MUSEUM, IWM CS 33. PHOTOGRAPHER: WOODBINE (F/O).

First fighter squadron, preparing to go overseas.

PHOTO: ANONBY FAMILY COLLECTION.

By the end of June 1941, some 280 airmen and officers had been posted overseas, including most of the whalers.[12]

Another squadron of Army Air Service personnel from Little Norway went to Catterick, England, in November 1941. They joined the RAF's 332 Norwegian Squadron, formed on January 16, 1942. Both 331 and 332 were distinctive Norwegian units under the operational command of the RAF's 13 Group, Fighter Command. The two squadrons served as escorts for RAF bombers.[13]

A change of command

On March 28, 1941, King Haakon decreed that the Navy and Army Air Services would merge to become one body, the Royal Norwegian Air Force (RNAF). Now, instead of two separate chains of command at Little Norway, there would be one. However, the two Air Services did not officially amalgamate until November 1944.

Captain Hjalmar Riiser-Larsen, formerly commander of the Navy Air Service, became commander in chief of the RNAF at its headquarters in London, England. Major Bjarne Øen, former commander of the Army Air Service, was made Risser-Larsen's chief of staff.

Effective April 16, 1941, Lieuteant Colonel Ole Imerslun Reistad took control at Little Norway as commanding officer. He had represented his country at the 1920 Olympic Summer Games in Antwerp, Belgium, and in 1922, been Norway's champion in pentathlon. That same year, Reistad had been honoured with the celebrated *Egebergs Ærespris* award, Norway's foremost prize in sports, presented for achievements in multiple sports. In the 1928 Olympic Winter Games in St. Moritz, Switzerland, Reistad had carried Norway's flag at the opening ceremonies and led his team to win gold in the military patrol event, a forerunner of the biathlon. He was also a skilled aviator, test pilot, flight instructor, communicator, and natural leader.

Reistad had escaped from Norway, after the country surrendered, by travelling through Sweden, Finland, and the Soviet Union to Japan. From Japan, Reistad signed onto a ship heading for Vancouver, working as a seaman on board the vessel; leaving Vancouver, he took a train to Toronto, arriving early in December 1940.[14]

Per Conradi Hansen used these words to describe Ole Reistad:

> Intelligent and highly articulate, he was a master motivator, who not only in spirit, but also in his personal behavior and deeds, set an example that inspired all of us who came to "Little Norway." ... Most of all, he was "one of us," the highest accolade the air men of "Little Norway" could bestow on an officer. [15]

In a press conference held in May to announce the RNAF's changes in command, Reistad strove to squelch some rumours that had been spreading about the Norwegian airmen. Although he didn't specify what these rumours might be, it seems that the free-spending ways of some of the Norwegian trainees had concerned Toronto citizens, who wondered if the Nazis might be financing the training camp.

Reistad told the press that these rumours were simply not true and had been instigated by fifth columnists attempting to undermine the work of the Norwegians. He told reporters that the recruits received a careful vetting by Norwegian authorities before coming to Little Norway, and that there was close contact with the Royal Canadian Mounted Police in these matters. He stated, as had been explained by Norwegian authorities time and again, that all the money required to run the camp came from Norway's merchant marine.

On a lighter note, Reistad thanked Canadians for the hospitality they had shown to the Norwegian airmen and remarked that he'd received a number of applications from his personnel to marry Canadian girls. He warned:

> I wish to ask Canadian women to think the matter over well before contemplating marriage with a Norwegian. Their love must be prepared to stand many trials. Our boys are at war, and will have to go overseas at any time. We do not know how long the Norwegian Government will be in a position to pay wages and pensions. When we return to Norway, we go to a country on the verge of ruin. The standard of living will be so reduced that one must be a Norwegian to stand the strain. We shall be able to because we shall be home again and are accustomed to a rigorous code of life.[16]

However, Major Reistad's advice generally fell on deaf ears. Young women continued to arrive at Little Norway's main gate after 5 p.m. each day to meet their boyfriends. Over 10 per cent of the airmen who trained at Little Norway would marry Canadian women.[17]

The newly married women often wanted to learn to speak Norwegian. Volunteers like Mrs. T. Berg, President of the Ladies' Auxiliary of the Norwegian War Aid Fund Association of Ontario, gave language lessons to the war brides and offered advice on their upcoming lives in Norway. She said, "I tell them they'll be homesick when they first go to Norway—that it will take a couple of years to become adjusted to new ways of living and making new friends." [18]

Knitters and Vikettes

A group of officers' wives, headquarters' staff, and some local Norwegian women began to meet regularly in early 1942, to assist the RNAF and Norwegian Relief by knitting mittens, sweaters, tuques, and socks. In fact, much of their time was spent mending socks for the airmen.

At first, meetings were held one evening a week at the tailoring shop at Little Norway. Later, the sessions moved to rooms on Gould Street and, finally, in the spring of 1943, relocated to the RNAF's equipment depot on Lombard Street.[19]

Rita (Abel) Hiorth of Toronto, who was about 15 years old

A group of officers' wives, Royal Norwegian Air Force staff, and local Norwegian women met regularly to mend the airmen's socks and knit woolen goods. Front row, third from left with dark curly hair, is Gunda Strom, Elsa Strom's mother. Front row, fifth from the left, is Janna Ullmann, actress Liv Ullmann's mother. Moving to the right side of the table, front row, first from the right side of the photo, is Rita (Abel) Hiorth. Rita's mother, Pauline Abel, is sixth from the right wearing a printed dress. PHOTO: GRAVENHURST ARCHIVES.

at the time, accompanied her mother, Pauline Abel, to the group mending sessions. There she found large cartons of freshly laundered socks, some with holes as large as oranges.[20]

Elsa Strom and her mother also belonged to the knitting group. Elsa had been born in Trondheim and came to Canada as a young child with her parents. Because she could speak both Norwegian and English, Elsa later worked on the switchboard at Little Norway.[21]

Other members of the knitting group included the mother and grandmother of Norwegian actress, Liv Ullmann. Lieutenant Erik Viggo Ullmann, Liv's father, was an aircraft engineer who had been working in Tokyo when Germany invaded Norway. He came to Canada to join the RNAF, bringing his wife, Janna, his mother, Dagmar, and his two small daughters, Bitten and Liv. The family first lived on Centre Island and later at St. Leonard's Avenue in Lawrence Park. Liv, who was two years old at the time, spent the next four years in Toronto and attended kindergarten in the city. Then, her father suffered a terrible mishap.

Lieutenant Ullmann accidentally walked into an airplane's spinning propeller and was seriously injured. He seemed to recover well at first, and was assigned to a new job in New York City, but when he grew weaker and weaker, doctors discovered he had a brain tumour. Lieutenant Ullmann underwent several brain operations but passed away in 1945.

Liv Ullmann told a *Toronto Star* interviewer during a visit to Toronto, "I was here during the worst and the best times of my life."[22]

Members of the Vikettes Club entertain at a Variety Show, Toronto, January 16, 1945. From left to right: Edna Walker, Mabel Young, Mabel Hatfield, Asa Thessman, Billie Oakes, Joy Brooke, Betty Holme, Edna Jones, Karen Larson, Ethel Anderson, Bunnie Larson, Doris Quittenton, Ruth Lavery, Florine Ketcheson, Gale Cooke, and Bonnie Winter. PHOTO: LITTLE NORWAY MEMORIAL

Another group of women, called the "Vikettes," also volunteered their time for the RNAF. They tended to be younger in age than those in the knitting group and while the Vikettes also mended socks and knit clothing items, their focus was directed towards organizing entertainment, parties, and dances.

15

Resistance Grows

Within occupied Norway during 1941, civilian resistance became more structured with the creation of the Military Organization or "Milorg." The organization's central command provided direction to five loose, regional groups of fighting patriots. Milorg's main goal was to build volunteer guerilla forces so they could assist with a future Allied invasion of Norway; in the meantime, the organization wanted its members to keep out of the spotlight, eschew sabotage, and minimize the chances of German attack. Its motto was "lie low, go slow."[1]

Late in 1941, King Haakon and the Government-in-exile formally recognized Milorg as a branch of the Norwegian Armed Forces.

Milorg's "lie low, go slow" view was directly contrary to the approach taken by the British Special Operations Executive (SOE), which favoured acts of sabotage, such as a March 1941 hit-and-run raid on the Lofoten Islands, an archipelago in Norway. British and Norwegians soldiers carried out the attack, including a few Norwegian commandos from Norwegian Independent Company No. 1. During the raid, the invaders sank some ships that were under German control and destroyed a fish-oil plant: fish oil was useful to the Germans in making explosives. They arrested some 200 German sympathizers and brought them back to England, along with over 300 Norwegian patriots. Some of these patriots made their way to Canada and joined the RNAF.

However, Josef Terboven, the Reichskommissar in Norway, took his revenge on the Norwegian patriots who had left for England with the invaders; he ordered that their property be burned. Police arrested individuals who had assisted the British during the attack, and the island population, as a whole, received a fine.[2]

Sometimes patriots played a double role, serving as members of the Norwegian police or military while assisting the Resistance at night. It could be dangerous and frustrating work. One of these men was Rolf Adolf Hansen, who lived in Porsgrunn. During the day, he was a Norwegian soldier, but at night, he secretly assisted with acts of sabotage.

One day, Hansen's frustration with the German occupiers overcame him. Putting on his uniform, Hansen told his wife he was going to shoot the first German he saw. He subsequently got into a fight with a German soldier, shooting and injuring him. During the struggle, Hansen lost a button from his uniform. The Gestapo used the button to trace him and put him under arrest. Rolf Hansen received the death penalty; he was the third Norwegian executed by the Germans. His brother Alf Hansen escaped to join the RNAF.[3]

As Terboven and the Quisling Government attempted to tighten control over Norwegians during 1941, many organizations fought back.

The bishops of Norway's state church, the Lutheran Church,

united to write a "pastoral letter." The letter gave three indicators that the justice system was disintegrating: first, the Supreme Court had resigned; second, acts of violence were being encouraged by the Nasjonal Samling; and, third, clergy were being coerced to betray confidences of their parishoners.[4] News of this letter spread internationally after ministers read it to parishioners during Sunday services across the country.

The Nasjonal Samling asked every teacher to sign an oath of loyalty to the new order, promising to promote pupils' understanding of the government's ideology and program. However, the teachers' organizations drafted a standard letter of response for their members to sign. Rather than pledging loyalty to Quisling's regime, the teachers pledged to be faithful to their calling and conscience.[5] As a result, the Nasjonal Samling backed off for a time.

Terboven wished to have voluntary associations, such as trade, professional, sports, and social organizations, adhere to the new order. In April 1941, 22 organizations of civil servants and connected occupations wrote a common letter of protest to Terboven; this was followed the next month by a similar letter signed by 43 organizations. As a result, the Gestapo arrested nine of the signatories, and the Minister of the Interior dissolved many of the associations. Members and officials of organizations like those for dentists, lawyers, and engineers resigned from the bodies en masse, though the associations continued to exist underground.[6]

Quisling set up a "Nordland" army regiment and appealed to Norwegian youth to join and fight "shoulder to shoulder" with Germany. However, few volunteers signed up.

Food shortages became commonplace in Norway; meat, butter, eggs, and coffee, among other items, were scarce.

Alf Hansen

PHOTO: ERIK HANSEN COLLECTION.

When it appeared that the Nasjonal Samling was about to take away the daily milk rations of workers in Oslo factories, a spontaneous strike broke out on September 8 and 9, 1941.

On September 10, Terboven declared a state of emergency in Oslo. The Gestapo arrested many union leaders and, that same day, executed two of them, Viggo Hansteen and Rolf Wickstrøm. For the next six days, journalists, newspaper editors, university professors, and others considered enemies of the new regime were arrested. Finally, on September 16, Terboven lifted the state of emergency.

As a result of the September events, all wireless radio sets owned by Norwegians were ordered turned in to the authorities. This was a blow to the Resistance, as the BBC provided

its members with important news and messages. But without easy access to radio broadcasts, people turned to newspapers, pamphlets, and leaflets, which were passed from hand to hand. Some people circulated chain letters containing news, patriotic messages, and poems. In late 1941, about 20 newsletters with names like *We Want Our Own Country* and *The Royal Post* were being printed by means of duplicating machines. Readers were often asked to make their own copies and issue them to their friends.[7] But those who became involved in this process did so at grave risk. One who paid the ultimate price was Harald Hvidsten, a brother of RNAF member Per Hvidsten.

Hvidsten family members owned a printing plant and a left-leaning newspaper in Sarpsborg, Norway. Harald Hvidsten served as editor of the newspaper. When newspaper publishing came under suspicion during the German occupation, the family ceased business. The final edition of their newspaper was printed on October 1, 1941. Although Hvidsten was kept under surveillance by the German police, he secretly published underground newspapers at the family's printing plant for a short time until he was arrested. Hvidsten was sent to a prison in Germany, where he died in 1943.[8]

TOP Military parade in Oslo, November 1942. From left, Valentin Feuerstein, Hermann Boehm, Vidkun Quisling, and Josef Terboven.

PHOTO: NATIONAL ARCHIVES OF NORWAY.

BOTTOM Vidkun Quisling attended a Nasjonal Samling conference in Skien, Norway, March 1942. Here, he inspects a unit.

PHOTO: NATIONAL ARCHIVES OF NORWAY.

Inger Stene

Inger Dietrichson, from Oslo, was a 22-year-old secretary at an electrical company when Germany invaded Norway. She became involved with Resistance work, at first distributing illegal newspapers and guiding people sought by the Gestapo from one hiding place to another.

When the Germans requisitioned all wireless radio sets in late 1941, Inger's father, Harald, bought a cheap set which he then turned in to the German authorities. He hid their much better wireless set in the basement in a wood pile, behind certain pieces of wood he had specifically cut out for the purpose. Inger and her father would go to the basement to listen to wireless broadcasts from the BBC, which she would then type up and give to her Resistance contacts.

Hildur, Inger's mother, would sit in the kitchen above the basement during the radio broadcasts, holding a fork. If someone came up the driveway, Hildur's task was to hit the kitchen radiator with the fork loudly enough for Inger and Harald to hear the sound. They would scramble to put the wireless set back in the woodpile and place the cut pieces of wood in front of the radio. Luckily, they were never caught.

Inger also served as a weapons courier, carrying handguns and ammunition in suitcases from place to place. On one Sunday morning, she had been to the home of Harald Edelstam, the Swedish consul to Oslo, who was sympathetic to Norway's plight and took great personal risks in helping the Resistance. He was affectionately called "The Black Pimpernel." At Edelstam's home, Inger had picked up a heavy, weapons-filled suitcase.

As she stood alone on the train station platform, she noticed a young German officer suddenly appear. He paced back and forth. Inger didn't know whether he'd watched her enter and leave the Swedish consul's home, or had otherwise been following her. She felt him looking at her, and looking at the suitcase, and was deathly afraid he might politely ask, "Fräulein, may I help you with your suitcase?" If he tried to lift her suitcase, he would immediately suspect what it contained. To Inger's enormous relief, he did not approach her, and she was able to complete the weapons delivery.

During the period when Inger was working with the Resistance, she was known by a cover name, "Eva Smith-Peterson," and lived in many different places, but never at her real home. She called her mother daily, though, and limited the conversation to ordinary topics; that way, if the police were eavesdropping, they wouldn't learn anything useful.

One of Inger's main tasks for the Resistance was to listen in to Gestapo and police telephone lines and convey the information to her contacts. A patriot in the telephone company had set up the phone connection. When Inger and her friends learned the patriot had been discovered and arrested by the Gestapo, they knew they had to escape quickly from Norway.

Her Resistance contacts told Inger to take a train to a certain station, where she was met by a man who brought her to a safe place for the night. The next day, two men who were also trying to escape from Norway joined her. They all skied towards Sweden, stopping for the night at a log cabin along the route. No one had been living in the cabin for some time, except for some bed bugs—which bit Inger as she slept.

The following morning, the group continued on skis for some distance, but had to resort to trudging through the snow on foot because they thought the skis made too much noise. It was exhausting, but Inger could see the lights of Sweden in the distance and knew she had to keep going and get there. And she did.[9]

16

Sports

Many of the Norwegians were excellent skiers; as one remarked: "You have to be able to ski to get anywhere in Norway in the winter."[1]

During their first winter in Canada, some RNAF trainees spent several weekends in Huntsville, 230 kilometres north of Toronto, practising for an upcoming ski-jumping competition in the United States. Spectators were thrilled to watch Little Norway's Lieutenant Stein Sem, who soared off a 30-metre ski jump at a speed of 80 kilometres per hour, on one ski.[2]

A particularly challenging ski jump had been built in 1934 at Mica Mine Hill, an abandoned feldspar and mica mine located close to town. This ski jump rose some 83 metres above Fairy Lake, and the jump itself was 52 metres long. Jumpers couldn't see the bottom of the hill from the top and could travel at speeds of up to 113 kilometres per hour, often landing on the lake. After the jump, a skier had to walk back up to the top as no ski lifts existed.[3]

A Buffalo newspaper editor organized an "international ski weekend" in January 1941. A trainload of 50 skiers from Buffalo and 100 from Toronto, including 20 men from Little Norway, travelled to Huntsville for the occasion.

A three-piece band played from the time the train left Union Station, in Toronto. Before long, the baggage car was cleared so a barn dance could take place. Some passengers started sing-songs as they travelled along. "There'll Always Be an England" was a favourite with the Americans.

When the train arrived at the Huntsville station, hundreds of townspeople enthusiastically cheered for the visitors, and magnesium flares lit up the darkness.

Some of the skiers stayed in Huntsville, but the majority travelled by bus to Limberlost Lodge where they managed to get a few hours' sleep. Saturday brought skiing, snowshoeing, tobogganing, skeet shooting, skating, and riding. Twenty Norwegians, who were training in the area at the time, skied around 10 kilometres from a nearby lodge to greet the American visitors and join in the fun.[4] An *Evening Telegram* reporter wrote:

> Among the most popular guests of the week-end were the blue-clad men of the Norwegian Air Force who proved indefatigable in their determination to teach even the most unpromising American ski-tyro the gentle art of remaining upright. All they asked of their pupils was that they be pretty, and willing to learn. Most of them were both.[5]

On Sunday, the group returned by train to Toronto and Buffalo: the Americans declared it was the best weekend they'd had in a long time.

Huntsville's winter sports committee organized a military ski championship for the weekend of February 1 and 2, 1941, reportedly the first civilian-sponsored military ski tournament in Canada.[6] Competitors came from Camp Borden and Little Norway in Ontario; Trois Rivières and Montreal in Quebec; and Sweden. In addition, a lone Dutch skier from the Netherlands Army, stationed in Stratford, Ontario, joined in the tournament.

Over 150 spectators arrived at midnight Friday by a special train from Toronto. A party spirit prevailed in Huntsville, with many dancing to music at the town hall until the early morning hours. The town's hotels were all fully booked, so private citizens generously opened their homes to help deal with the overflow of guests.

As planes from Little Norway flew and swooped above to encourage their ski team on Saturday, the Norwegians took the honours in almost every contest. The exception was the downhill race, where Lieutenant C.F. Reiffenstein of the Canadian Grenadier Guards, Camp Borden, came out on top, with Norwegians in second and third positions.

Lieutenant Governor Albert Matthews drove to Huntsville Saturday morning from Toronto, and cut a red ribbon to officially open the ski-jumping competition at Mica Mine Hill. This competition was particularly exciting to watch and drew the biggest crowd, including Lady Flora Eaton and a party of her friends, who had travelled from Toronto to see the events. The *Toronto Daily Star* reported that the ski jumping thrilled Lady Eaton.[7]

Lieutenant Stein Sem won the ski-jumping competition with a jump of 40 metres; not satisfied, he tried again after the competition had finished, this time achieving 41.4 metres, just short of the all-time record. Other ski jumpers from Little Norway finished second and third in the contest.

Saturday night brought an awards ceremony and another big dance, during which one airman from Little Norway told the crowd over the loud speaker: "We like Canada, and we've never seen so many pretty girls in one place."[8]

Tug of war at Little Norway, Toronto. Egil Chramer, at the left side of the photo, pulls the rope with his team.

PHOTO: BARBARA CHRAMER COLLECTION.

Little Norway's Ulf Wormdal sped through the 16-kilometre cross-country ski race on Sunday morning, a course that rambled up hills, through swamps, and down steep rock faces. He came first; the next eight finishers were all from Little Norway.

Norwegian trainee Marius Eriksen won first place in slalom on Sunday afternoon.[9]

Track and field at Maple Leaf Stadium.
Sergeant Arne Skretteberg is second from left, with arms folded.

PHOTO: MICHAEL SKRETTEBERG COLLECTION.

The Norwegian airmen excelled at summer sports, too, including swimming and track and field. The RNAF created a soccer team to play local and military teams in charity games; the airmen usually beat their opponents. The Special Services Branch of No. 1 Training Command, RCAF, included the RNAF in all its inter-unit competitions.[10]

Track and field sports were hugely popular at Little Norway. In June 1942, the Sports Service League for Canadian military personnel organized a 16-kilometre relay race with a course running in a zigzag fashion through Toronto's hilly High Park. The RNAF entered two 15-member teams, coached by Sergeant Arne Skretteberg. There were a couple of other teams, including one from the RCAF's Manning Depot.

The Norwegian teams came in first and third in the relay race, with the RCAF team taking the second spot. Reportedly, one of the Norwegian teams nearly lost the race completely, when a relay team member decided he had time to take a short drive with his girlfriend while waiting for the preceding runner to reach him. He returned just in time to see the team member he was to relieve barrelling down the road. The airman got to his starting point just in time to jump out, grab the baton, and race off.[11]

Civic contributions

Little Norway actively supported patriotic drives to help Canada finance the war effort. One important initiative was the Canadian War Savings Campaign, which solicited pledges from the public to buy War Savings Certificates and Victory Bonds. Campaigns took place in various cities and towns across the country. Customarily, a large thermometer was placed at the front of a public building to demonstrate increases in the number of pledges made, its red indicator rising as pledges flowed in.

In Toronto, the campaign was officially launched on February 22, 1941, when over 100,000 people gathered to watch an eight-kilometre-long parade proceed along downtown streets to City Hall Square.

Parade-goers roared their approval as military vehicles and personnel passed by. Included in the parade were army trucks, field guns, Bren gun carriers, motorcycles, four American six-ton tanks, and a British 15-ton tank. Thirty-six military marching bands participated, as well as 10,000 representatives of the army, navy, and air force, including units from the RNAF. Brigadier-General G. Eric McCuaig took the salute from a reviewing stand at the corner of Yonge and College Streets. After the parade, tanks gathered at Fleet Street and demonstrated a mock tank battle, thrilling several thousand viewers.[12]

The campaign's pledge thermometer, located at city hall, rose to 103,000 as a result of 22,000 pledges given after the parade. By mid-March 1941, the campaign had reached its objective of 200,000 pledges.[13]

The RNAF also supported launches of the War Savings Campaign in smaller Ontario cities, when the pilots would fly aircraft in formation above the parades. In December 1940, several planes were to drop leaflets over Chatham, Ontario, promoting the purchase of War Savings stamps. The City of Chatham later wrote a letter to the Norwegians, thanking them for their "wonderful airmanship" and noting that they appeared over the city "exactly to the minute as arranged." This was a lucky turn of events, as the planes had met with strong head winds, which had blown them off course. One pilot, somewhat unclear on his Ontario geography, and believing his plane must be over Chatham, dropped 5,000 leaflets over Hamilton.[14]

"Spirit of Little Norway" aircraft fund

As wartime losses had reduced the size of Norway's merchant marine from 5,600,000 tons to 3,600,000 tons, less money was available to Norway's Government-in-exile to support its war effort, and that included purchasing training planes.[15]

On May 17, 1941, Norway's Constitution Day, a very special event occurred: in a "17th of May Greeting," as it was called, every member of the RNAF pledged a portion of his monthly pay for the duration of the war towards "The *Spirit of Little Norway* Fund," a savings account established for the purchase of new airplanes. New recruits arriving at Little Norway subsequently could join in the 17th of May Greeting by signing a pledge card.

The idea had originated among the airmen themselves, and, instantly, became very popular.

Contributions were proportionate to an individual's rank. For instance, a lieutenant donated $10 out of a monthly net salary of about $105; a sergeant, $5 out of a monthly net salary of $89; a private, $2 out of a net salary of about $30.[16] Additional voluntary donations, such as those made by former sailors with accumulated back pay, helped increase the fund.

By August 1941, the first *Spirit of Little Norway* airplane had been purchased. Over the course of the war, some $45,000 would be raised from the contributions of RNAF personnel, allowing two more Fairchild training planes to be purchased. The aircraft were named *Spirit of Little Norway* I, II, and III.[17]

A number of aircraft were also obtained through a funding drive originated by Swedish-Americans called "Wings for Norway," headquartered in New York City. Members of the board of directors included the American poet and writer Carl Sandburg, who wrote in a magazine produced by the organization:

> And we Americans of Swedish descent who are alive to history in the making before our eyes—we feel a call of blood. We cannot look with cold indifference on the present scene in the Scandinavian Peninsula.... This enterprise of Wings for Norway comes out of American soil and air and has no official connection with any government.... The answer now is to raise the funds that will guarantee the Norwegian youth the training planes they need. The big word is "Send those boys in Canada the training planes they are calling for."[18]

Committees formed all over the United States, organizing rallies, benefit concerts, and collections for the purchase of training aircraft for the RNAF in Canada.

Generous contributions also flowed from Scandinavians residing in South America, Mexico, and other countries. Norwegians living in Uruguay and Argentina donated a new Fairchild training plane, valued at $10,200, which arrived on March 27, 1941. The donors requested that the plane be called *El Gaucho* and they sent along a gaucho doll made in Argentina as a mascot.[19] There would be three more *El Gaucho* aircraft donated by this group during the war.

Norwegian-born Leif Lind Pettersen owned a coffee plantation in Guatemala and donated coffee to the Norwegian Armed Forces in Canada. In January 1942, Petersen contributed a Fairchild aircraft to the RNAF. The aircraft sported a painting of a steaming coffee pot.[20]

Other donations at that time included the *Minnesota*, funded by Norwegian-Americans from that state, and the *March of Time*, purchased with the proceeds of a film made about the Norwegian occupation.[21] Over the war, more aircraft were donated to the RNAF, including the *State of New York*, by Norwegian-Americans, and *Denmark* I and II, by Danish-Americans.

Publicity

Lieutenant Colonel Ole Reistad wanted to publicize Little Norway in the United States to promote fundraising for purchasing aircraft. He sent 10 veterans of the war in Norway on a goodwill mission to Washington, DC, flying five Douglas attack bombers. The group included Captain John Stene, Lieutenant Per Waaler, and Lieutenant Rolf Tradin. They left Toronto on September 25, 1941, circling the White House and Capitol a couple of hours later, and subsequently meeting with Crown Princess Märtha in New York City.[22]

Another idea aimed at drawing attention to Little Norway's needs was the creation of a postage stamp. Members of the American military forces serving in Canada were already allowed to use United States' stamps to send their mail; Lieutenant Colonel Reistad thought that if there was a Little Norway stamp, Canadian authorities might allow the Norwegians to do the same thing.

He made arrangements with the Canadian Bank Note Company to print a stamp based on a "Wings for Norway" magazine cover designed by Captain Harald Nicolaysen, Chief of Little Norway's Radio School. George Unger Vetlesen, a Norwegian-American businessman, helped finance the venture.

The stamp shows two airmen walking on top of the globe, one standing in Canada and the other in Norway, with a ghostly Viking behind them. The man on the right side of the stamp is Sergeant Ulf Wormdal, who had been a professional illustrator before the war and was a star member of Little Norway's ski team. On the left is Quartermaster Henry Bernhard Malmoe.

Both young men would later be killed at age 24. Wormdal died on manoeuvres with Squadron 331 in the Orkney Islands during 1942; Malmoe was lost in action while flying a patrol flight from Scotland in 1943.

Although Lieutenant Colonel Reistad made persistent efforts to have the stamp issued, the Canada Post Office did not do so, as it never received a formal request from Norway's Government-in-exile on behalf of Little Norway. It has been speculated that as the Government-in-exile was itself planning to issue a set of stamps in England for use by offices abroad, it wanted to avoid the confusion of having two different sets of stamps in circulation.

After the war, Norway's Government issued a stamp with the same design as that created by Captain Nicolaysen, to demonstrate Norway's appreciation for Canadian assistance during wartime and to honour those who served at Little Norway.[23]

Stamp with likeness of Quartermaster Henry Bernhard Malmoe on the left and Sergeant Ulf Wormdal on the right.
PHOTO: MICHAEL SKRETTEBERG COLLECTION.

17

Growing Pains

The United States claimed neutrality at the beginning of World War II: most American citizens wanted to keep out of the war, which many viewed as a European conflict.

The American *Neutrality Act* prevented the United States from selling military supplies to a foreign country that was at war, unless that country paid for the items on a cash and carry basis. However, President Roosevelt wanted to provide war supplies to the United Kingdom, as requested by Winston Churchill, even though the United Kingdom couldn't afford to pay cash.

Roosevelt devised a solution to the problem, creating the "Lend-Lease Program." Under this program, the president could sell, lend, or lease military articles to any country whose defence the president deemed vital to the defence of the United States. Repayment could be deferred until the hostilities had ended. To avoid creating war debts, the eventual repayment didn't need to be made in cash; it could be some other type of consideration.

The United States allocated goods worth $1.6 billion to Norway's Government-in-exile under the Lend-Lease Program, for use by the Norwegian Armed Forces. By June 15, 1945, Little Norway had received $1,300,800 worth of supplies from that allocation, including 50 Fairchild PT-26 training aircraft, 32 motor vehicles, ammunition, clothing, 245,782 gallons of aviation gasoline, 4,520 gallons of oil, and 61,095 gallons of automobile gasoline.[1]

Troubles at Toronto Island Airport

By the spring of 1941, more than 800 personnel lived at Little Norway, and around 90 Norwegian aircraft operated from Toronto Island Airport.

With this rapid growth came increased expenses, such as a surge in the consumption of aviation oil and gas, and the RNAF began to examine its fuel costs. The September 1940 Agreement between Norway and the Toronto Harbour Commissioners required Little Norway to purchase all oil and gasoline needed for its operations from the Harbour Commissioners at retail prices.

The Harbour Commissioners had an arrangement with oil companies by which they purchased aviation gasoline for Toronto Island Airport at a discounted price. The Harbour Commissioners then set a retail price for the resale of the aviation gasoline to airport users, which included a profit for the Commissioners. Depending on the type of gasoline involved, the profit ranged from seven to seven and one-half cents per gallon. The same retail prices were charged to all who purchased aviation gasoline at Toronto Island Airport, whether they were from the RCAF, RNAF, or the private sector. With a view to cost savings, the RNAF wondered whether the Harbour Commissioners might lower their profit margin and so decrease the overall cost of aviation gasoline.

The RNAF chiefly used a type of aviation gasoline known as "90 Octane." Toronto Island Airport had no storage facilities for this type of gasoline, so RNAF staff had to drive their truck to the oil companies' bulk-storage facilities on the mainland, pick up the 90 Octane gasoline, drive back, and then refuel Little Norway's aircraft themselves. Even though the Norwegians had to take on this responsibility, the Harbour Commissioners allowed no discount from the full retail price.

As for aviation oil, the Harbour Commissioners paid $0.93 per gallon for the product and sold it to the RNAF for $1.40 per gallon.[2] To the Norwegians, this seemed to be a steep mark-up.

The RNAF asked the Harbour Commissioners to renegotiate the provision of the Agreement that required them to purchase gasoline and oil at the Commissioners' retail prices. Instead, the Norwegians wanted to buy their fuel directly from the oil companies.

The Harbour Commissioners refused to make any changes to the Agreement. They pointed out that although the City of Toronto did not own the Toronto Island Airport, the Harbour Commissioners oversaw its operations on behalf of citizens. The Harbour Commissioners argued that they had a duty to manage the airport as a profit-making enterprise for the city, even though the airport usually operated at a loss. The RNAF had use of nearly 90 per cent of a $1.25 million airport for next to no costs, the Harbour Commissioners said, and also reminded the Norwegians that they paid no rent for their camp's use of the lands at the foot of Bathurst Street. In summary, the Harbour Commissioners believed that the RNAF's complaints were totally unjustified.

Little Norway's machine shop. Odd Toldnes is on the left.

PHOTO: KAREN GREAVETTE COLLECTION.

As it had gotten nowhere with the Harbour Commissioners, the RNAF approached Toronto's mayor, Frederick Conboy, as well as members of Toronto's Board of Trade, Canada's Department of Munitions and Supply, and others. These

Classroom, Little Norway, Toronto. On the left is Alf R. Bjercke.

PHOTO: S.J. REGINALD SAUNDERS, *LITTLE NORWAY IN PICTURES.*

officials would write to the Harbour Commissioners, asking for an explanation of gasoline and oil pricing. In reply, they would receive a standard letter, explaining the Harbour Commissioners' position and reiterating that the Agreement was not going to be renegotiated.[3] That was usually the end of the matter.

In fairness to the Harbour Commissioners, they were very aware of their duty to manage the lands as trustees for the city and wished to obtain the best possible financial outcome for Toronto's citizens. Naturally, this approach often came into conflict with the RNAF's obligation to spend the Government-in-exile's capital economically, carefully, and efficiently.

The RNAF did receive one small concession regarding fuel costs from the oil companies themselves, negotiated through Canada's Department of Munitions and Supplies. The oil companies agreed to give the Norwegians a discount on the price of gasoline equal to that given to the British Commonwealth Air Training Plan Schools.

On May 21, 1941, the Harbour Commissioners met with a delegation from Little Norway, led by Lieutenant Colonel Reistad and the RNAF's lawyer, William Zimmerman, to discuss the various issues of concern.

The Norwegians asked to be allowed to buy fuel directly from the oil companies rather than having to purchase their oil and gasoline from the Harbour Commissioners.

They also raised concerns about the non-military use of Toronto Island Airport. After the RNAF began to use Toronto Island Airport for air-training purposes, the Harbour Commissioners continued to allow civilian flight schools to operate at the airport. RNAF training courses were constantly being interrupted by the flight schools' planes taking off, landing, and refuelling. In addition, many privately owned aircraft made use of the airport, coming and going from the airfield as they pleased, often interrupting and delaying the Norwegians' training.

The presence of visitors walking freely around the airport area raised security issues for the RNAF. As Lieutenant Colonel Reistad pointed out, a risk of sabotage always existed, and it would be very easy for someone to stealthily pour sugar in a plane's gasoline and contaminate it.

The RNAF occupied most of one hangar at Toronto Island Airport, using the space to assemble Curtiss Hawks and to repair aircraft. That left limited hangar space for storing airplanes. The Harbour Commissioners gave storage-space priority to the RCAF and civilian aircraft. As a result, the Norwegians' Douglas and Curtiss Hawk planes had to be tied down outside during the winter, which, as the RNAF representatives noted, was very hard on the aircraft.

Loose gravel, which flew up from the runway into the planes' propellers, was another problem, as was the difficulty of transporting large numbers of personnel expeditiously by ferry to Toronto Island Airport.

The Norwegians noted that they had invested nearly $300,000 in erecting buildings at Little Norway and Toronto Island Airport and, as a result, were hesitant to move elsewhere. However, something had to be done to remedy the various problems at the airport. The RNAF offered a solution: they asked the Harbour Commissioners to lease the Toronto Island Airport to Norway, for a nominal rent that would cover the Harbour Commissioners' expenses. The RNAF would operate the facility as a military airport for the duration of the war, under supervision of the Canadian Department of Transport if need be, and would return the airport to the Harbour Commissioners in first-class condition after hostilities ended.[4]

After the meeting, the Harbour Commissioners consulted with representatives of Canada's Department of Transport and talked further with the Norwegians, while formulating a response. In the end, the Harbour Commissioners decided that they would not lease the Toronto Island Airport to the RNAF and they would not change the Agreement requiring the RNAF to purchase gasoline and oil from them.

However, the Harbour Commissioners put forward a counter-offer: they would give the RNAF exclusive use of the airport, subject to the needs of the RCAF and Canada's Department of Transport. No civilian flight schools would be allowed to operate at the airport for the rest of the war and the contract between the Harbour Commissioners and the one flight school currently instructing students at Toronto Island Airport would be terminated.

More hangar space would be made available to the RNAF, as civilian aircraft would no longer be stored at the airport. Some other minor matters would be remedied, such as removing loose gravel from the runways and streamlining the ticketing process for ferry transport.

The RNAF accepted the Harbour Commissioners' counter-offer but reserved the right to discuss the question of reducing airport costs and charges at a later time.[5]

Beginning in the fall of 1942, the United States' Lend-Lease Program's allocation of oil and aviation gasoline assisted the RNAF. Under this program, the oil companies billed the Navy Department in Washington directly for fuel supplied to the RNAF. However, the Norwegians still had to pay the Toronto Harbour Commissioners their profit on the fuel sales as required by the September 1940 Agreement.

More accidents

Whatever happened to Second Lieutenant Kjell Christensen on April 18, 1941, remains a mystery. He had been carrying out a high-altitude test flight in a Curtiss Hawk aircraft, taking the plane to an altitude of some eight kilometres above the earth, when he disappeared. Neither Lieutenant Christensen nor his aircraft were ever found.

Six witnesses near Mattawa, Ontario, located about 400 kilometres north of Toronto, reported seeing a plane flying very low over the treetops on the morning of April 18, which was an unusual enough event that they clearly remembered it. However, extensive searches by aircraft from the RNAF and RCAF over the Mattawa area for 10 days failed to locate any evidence of the Curtiss Hawk or its pilot.

A search plane from Little Norway reported seeing an oil patch in Lake Ontario near Burlington, and Norwegian offic-

Airmen in shorts.

PHOTO: LITTLE NORWAY MEMORIAL.

ers suspected that Lieutenant Christensen had crashed into Lake Ontario.[6]

Lieutenant Harald Olafson remembered Christensen's Canadian fiancée coming to Toronto Island Airport day after day, desperately waiting for news. He felt badly for her and finally answered her questions about the fate of her fiancé, with the suggestion that the pilot probably drowned in Lake Ontario, sinking with his plane. To ease the blow, he added that drowning was said to be a pleasant death. The girl burst into tears, and never returned to the airport again.[7]

On April 26, an aircraft flown by Second Lieutenant Hans Petter Schwencke and Airman Bjarne Nilsen crashed near North Bay while making a right turn at a low altitude.[8] Both men died.

Another loss for the RNAF occurred on May 21, 1941, when a Harvard training plane flown by Corporal Aage Nygaard crashed near Moose Jaw, Saskatchewan.

Emsdale

Given the RNAF's concerns about the number of aircraft using the Toronto Island Airport, the Elementary Flying Training School moved to an airfield in Emsdale during the summer of 1941.

Emsdale is situated some 255 kilometres north of Toronto. During the Depression, the federal government had built an auxiliary airfield in Emsdale. The airfield was part of the Trans-Canada Airway System, an initiative intended to create a chain of airports and airfields linking the country from coast to coast. The Emsdale airfield was about the size of Toronto Island Airport, though rarely used.[9]

At the Emsdale camp, the student pilots practised takeoffs and landings in 15 Fairchild training planes. Lieutenant John Stene instructed a class there and he recalled: "It was a fantastic sight to fly in bright sun over the Canadian maple forests in their intense golden-red autumn colours."[10]

One student pilot who learned to fly at Emsdale was Rolf Kolling. After six or seven hours of flight training with an instructor present in the aircraft, a student could take a Fairchild up by himself. Lieutenant Stene sent Kolling off on his first solo flight on October 9, 1941. The flight lasted two minutes. In any pilot's life, the first solo flight is a huge step and Captain Kolling remembered it well: taking off in the Fairchild, circling around the airfield, and landing it safely.[11]

In a large hangar at the Emsdale airfield, mechanics maintained the aircraft, and student pilots learned to fly by instruments in a flight simulator known as a "Link Trainer." Emsdale also offered the trainees a shooting range for revolver practice.

Trainees lived in big tents; their instructors roomed at a nearby farmhouse.[12] When the men's out-of-town girlfriends visited them, the women often stayed in rooms rented from a village widow.[13]

An Emsdale reporter wrote in *The Huntsville Forester*, "The Norwegian airmen are back at the airport, and the hum of their planes makes us realize we are at war."[14] Another reporter noted that quite a few local people had been driving to Emsdale to watch the airmen practising landings, rolls, spins, and general flying practice.[15]

However, Emsdale proved inefficient for long-term use as a flying school, as it lacked equipment for night flying training. It also had space limitations and was located too far from Little Norway in Toronto. The RNAF's use of Emsdale terminated in May 1942.

18

Vesle Skaugum

Lieutenant Colonel Ole Reistad was worried about the temptations of Toronto.

He wasn't pleased to see so many of the young airmen spending their spare time in Toronto restaurants, bars, and dance halls. He wanted a healthier environment, closer to nature, where skiing, hiking, swimming, and other athletic endeavours would occupy the recruits' leisure time.

Reistad found a solution in an old fur trading post, 26 kilometres northeast of Huntsville, built on a portage between Oxbow and Long (Dotty) Lakes and known as "Interlaken." Settler Frank Peelar had obtained a Crown Grant to the lands in 1912. Gordon Hill, owner of Limberlost Lodge nearby, had later purchased the property from the Peelar family and used it as an outpost camp for Limberlost's horse-riding and cross-country-skiing guests. Hill had expanded the original cabin and added two or three log buildings and a stable.[1]

The Norwegians bought the property and surrounding lands, totalling some 175 hectares, in the spring of 1941 at a cost of $10,000. Funding came from contributions by Little Norway personnel, along with a personal donation, reportedly from coffee-plantation owner Leif Lind Pettersen. The RNAF paid ongoing operating expenses.[2]

The new Muskoka camp was named "Vesle Skaugum," in honour of the Crown Prince's and Crown Princess's home near Oslo, at the time occupied by Reichskommissar Josef Terboven. "Vesle" means "little" and so the camp's name meant "Little Skaugum."

Building Vesle Skaugum

Twenty-six-year-old John Engh had been an architect in Oslo before the war. After the invasion, he fought with the Norwegian Army, blowing up bridges and dynamiting roads to halt the advance of the German troops. He was captured by the Germans and sent to a prison camp in Norway, from which he managed to escape. As he'd been born in the United States before his parents returned to Norway, Engh obtained citizenship papers from the American Consulate in Stockholm and travelled to the United States. He soon joined the RNAF.[3]

Engh was retained to build a central dining, sleeping, and recreation lodge for the camp using traditional Norwegian log construction techniques. He told a reporter that Lieutenant Colonel Reistad wanted the buildings to have a connec-

OPPOSITE Aerial view of Vesle Skaugum between Long (Dotty) Lake in the foreground and Oxbow Lake to the rear.

PHOTO: LITTLE NORWAY MEMORIAL.

The main lodge at Vesle Skaugum, built with rounded logs and designed by John Engh.

PHOTO: ANONBY FAMILY COLLECTION.

tion to Norway in design; however, the budget was tight and the log buildings couldn't cost more than normal barracks.[4]

The architect decided not to follow traditional Norwegian methods in building the lodge, as that would have taken too long. For example, he and the Finnish carpenters left the peeled logs that formed the walls in their natural, round shape instead of hewing the logs to flatten them.

Timber for construction came directly from the woods surrounding Vesle Skaugum. One RNAF trainee, Johannes Helland, recalled being sent to Vesle Skaugum in early December 1941, while waiting to take the next course at Little Norway. John Engh was at the site at the time and he put Helland and other trainees to work in the woods, helping some Finns and Norwegian civilians cut down big trees and remove their bark. The men lived in a hut near the big lodge until around April. Helland recalled:

> It was marvellous here, it was ... cold but that was at night. We were working the whole day and when we were finished we were skiing the whole time. We were [skiing] every evening, even though it was more than 30 below.[5]

Thanks to Engh's designs and the builders' craftsmanship, Vesle Skaugum gained a beautiful, warm, and relaxing lodge, which housed a mess hall, sleeping quarters, and a living room. In the living room, wooden light holders were carved in the shape of Viking torches.[6] This building, along with accommodation provided by several small cabins, allowed up to 150 personnel to reside at the camp.

Per Ole Falkendal Nilsen from Aalesund, Norway, who was training to be a flight mechanic at Little Norway, remembered Vesle Skaugum in summer as being very much like Norway—but with bigger mosquitoes.[7]

On January 18, 1942, Crown Prince Olav and Crown Princess Märtha officially opened the main building, now completed with the Norwegian emblem posted over the doorway. Märtha traced a motto inscribed on a beam that had been built into the fireplace: "Merkje—Det—Stend—Um—Mannen—Han—Stupe," which loosely translated means, "The mark of man will stand though man may fall."[8]

The Royal couple told reporters that the camp, and the forests surrounding it, made them homesick for Norway.[9]

Lieutenant Erik Bjørneby was commanding officer at Vesle Skaugum. He had survived the March 1941 plane crash

Airmen relax beside the fire in an old log building at Vesle Skaugum. Limberlost Lodge used the property as an outpost camp for its guests before the Norwegians purchased the lands. Notice that flattened logs were used to construct the house, as compared to the rounded logs used by the Norwegians to build the main lodge. PHOTO: LITTLE NORWAY MEMORIAL.

Ski troops, Vesle Skaugum.

PHOTO: LITTLE NORWAY MEMORIAL.

in Patricia Bay, which had killed his friends, Lieutenant Kaare Kjos and Sub-lieutenant Jens Riiser. He had made a remarkable recovery: though Bjørneby had broken his back in the accident, he was able to stand straight and even ski when Vesle Skaugum opened.[10]

From the start, Vesle Skaugum served a variety of purposes. Those who had made dangerous escapes from Norway, had served on torpedoed merchant marine ships, or had suffered aircraft training accidents, could convalesce and recuperate at the camp. Little Norway personnel on leave could enjoy recreation in a beautiful location. And air crew who had finished training and were awaiting posting overseas could spend time there, working for half the day at the camp and relaxing for the other half.

Recruit School at Vesle Skaugum

In the spring of 1942, Vesle Skaugum took on an additional function: recruit training. As well as drills and physical training, the recruits learned commando skills and practised shooting at the rifle range. Recruit training lasted about 30 days.

During the winter months each year, recruits took part in a military ski competition, as cross-country-skiing ability was thought to be crucial to successfully waging warfare in Norway's mountainous areas. "This is no sporting event," Lieutenant Bjørneby explained to a reporter. "It is part of the regular training, which makes every air force man an all-round fighter on the ground, capable of defending and taking care of himself under any conditions he may meet," he said.[11]

On March 8, 1942, some 90 trainees and officers, including Lieutenant Colonel Ole Reistad, who was then 43 years old, set off on skis from Vesle Skaugum for a gruelling 32-kilometre cross-country ski exercise. They raced up and down steep hills and across frozen lakes in the terrain between Limberlost Lodge and Vesle Skaugum, with heavy rifles and nine-kilogram packs strapped to their backs. The rain drizzled down all day, adding about a kilogram in weight to their packs, and making the snow sticky.

At Limberlost, the competitors stopped at a food station offering fresh fruit, warm water with lemon, and cups of rolled oats mixed with raspberry juice, all designed to ener-

Vesle Skaugum in winter, blanketed in snow. Front view of an old log building on the property.

PHOTO: LITTLE NORWAY MEMORIAL.

gize them for the next part of the race. The men only spent about two minutes at the food station as their running time continued to be tallied during the break.

Next stop was a rifle range at Long Lake, where each fired 10 shots at a target. Lieutenant Colonel Reistad hit the target every time.

Reistad completed the punishing course in 3 hours, 13 minutes, and 14 seconds, faster than everyone in the entire competition.[12]

After finishing Recruit School in during 1942, 35 young Norwegians marched all the way from Vesle Skaugum to Little Norway in Toronto, some 250 kilometres. It took a week.

The group carried full equipment and camped out at night, cooking their own hot suppers, except on one occasion when they stopped in Orillia, and the citizens prepared them a feast.[13]

While Vesle Skaugum was very popular with many Little Norway recruits, Per Conradi Hansen was glad to finish Recruit School there. He recalled rising at six in the morning, undergoing a half hour of energetic calisthenics, then sprinting around the camp in the snow, wearing only a sleeveless shirt and shorts. After a shave, shower, and breakfast, the recruits gathered for formation and drill on skis carrying a full pack and rifle. Next came cross-country-ski racing, again bringing a full pack and rifle. "In the beginning, and this is no exaggeration, I felt after half an hour or so that I would die from a heart attack if I took another step," he wrote. However, after six weeks of this training, Hansen believed he was in the best shape he'd ever been.[14]

Regardless of how hard the recruits trained during the day, they still managed to have energy for evening activities. Men of all ranks mixed freely at the camp, playing soccer, chess, and, in the summer, swimming and canoeing. Airmen from Vesle Skaugum continued to take part in ski-jump contests during local wintertime celebrations. When they did, the Norwegians invariably won. One young trainee, Ola Aanjesen, became the 1942 United States National Ski-Jumping Champion, beating famed American champion Torger Tokle.[15]

On weekends, Huntsville welcomed the Norwegians, and they enjoyed themselves at the town's restaurants and at the popular pub at the Empire Hotel. Just as had occurred in the city, some trainees began dating local women. The beautiful setting itself inspired romance. Harald Jensen remembered one Midsummer's Eve, when a traditional Norwegian bonfire was built down by the lake, and girls were invited, both from local towns and from Toronto.[16]

Prince Harald takes a dip

Crown Princess Märtha brought her three children, Princess Ragnhild, 12 years old, Princess Astrid, 10 years old, and Prince Harald, 5 years old, to visit the Royal Norwegian Air Force in September 1942. They spent a few days at Vesle Skaugum. One morning around 7:00 a.m., while frost was on the ground and the temperature was below freezing, Prince Harald was observed dashing from the main building in a bathing suit, running to the lake, and diving into the cold water. He swam out to a diving tower, climbed to the top, jumped off, and swam back to shore. Harald repeated his fearless swimming adventure every morning during his stay at Vesle Skaugum, to the amusement and admiration of the airmen.[17]

19

Toronto Ferry Accident

Devastating plane crashes continued to take young Norwegian lives. By the time Vesle Skaugum opened in January 1942, 11 airmen had been killed.

The accident with the greatest reverberation occurred on June 20, 1941, when a Northrop floatplane crashed into the *Sam McBride*, a Toronto passenger ferry, as the boat was making its way across Toronto Harbour.

The *Sam McBride*'s captain, Ernest Straker, reported that the ferry was about one-third of the way across the harbour, heading to the city from Centre Island, when he saw the plane taking off from the water a half-mile away. It was heading straight for the ferry's bow, and though he tried to swing the boat away to avoid the accident, he couldn't. He said: "It was all over in a twinkling. The boat was going at 12 or 13 knots and the plane somewhere between 65 and 80 miles per hour. The impact was terrific."[1]

People riding the ferry watched in horror. "The plane headed right for the ferry, and I thought we were all gone. ... Suddenly, the plane zoomed up a little and struck part of the ship overhead," one passenger reported.[2] Some aboard speculated that by swinging the plane off-course at the last minute, the pilot was attempting to limit any harm to all those gathered on the boat's lower level.

The aircraft hit the upper level of the ferry and smashed a lifeboat, ripped off the roof of a pilot house, bent the mast, and left other damage. Fragments of the plane were found on the ferry's upper deck afterwards.

Lieutenant Finn Kjos and Airman Trond Harsvik died as a result of the collision, but not one of the 35 passengers on the *Sam McBride* ferry was harmed. Lieutenant Kjos's brother, Kaare, had been killed about four months earlier when his Northrop aircraft crashed in Patricia Bay, British Columbia.

When the plane's wreckage was raised from the harbour, the bodies were recovered.

An inquest took place on July 12, 1941. The chief coroner, Dr. Smirle Lawson, told the jury that crashes of RNAF planes were becoming quite frequent. He said: "It would be a miracle if some day these inexperienced pilots don't crash into a Toronto street." He also suggested the jury might consider recommending that the RNAF pilots train elsewhere.[3]

Lieutenant Commander Haakon Jøergensen, the officer in charge of the seaplane base, testified that every RNAF member was taught to practise safety at all times. Jøergensen said he had known Lieutenant Kjos very well, and that Kjos had been a very experienced and clever pilot. He felt certain that Kjos wouldn't have permitted the plane to start if he had seen a boat in its direct path.

Under cross-examination, Jøergensen admitted he was aware of complaints from Torontonians about low-flying, but said that north winds made low take-offs necessary, and the

Canadian air regulations allowed for them. However, knowing citizens' concerns, the pilots had made changes. Jøergensen added, "There seem to be many people in Toronto who do not realize we are preparing for war. This accident was one of those rare things which would probably not happen again in 100 years. It is certain there will be no such accident again." [4]

Other witnesses from Little Norway suggested that the plane had started up in the direction of the sun, and the pilots may have been temporarily blinded by its reflection from the water.

The jury decided that the two young airmen died by drowning, and that their plane crashed for reasons unknown. They declared their appreciation, as Canadians, for the part the Norwegian flyers were playing in the war. They then added their strong recommendation that the Norwegian Flying School and the Toronto Harbour Commission take definite steps to ensure the safety of Toronto citizens and the Norwegian pilots. The jury said: "We cannot emphasize too greatly the hazards involved when inexperienced flyers are manoeuvring in and around Toronto Bay and its environs." [5]

On July 14, the *Evening Telegram* contained an editorial supporting the chief coroner's views and noting a general concern that too much low-flying was occurring over the city's rooftops.[6]

Just after the deaths of Lieutenant Kjos and Airman Harsvik, two more young Norwegian airmen died in a fiery plane crash. On July 5, 1941, Sergeant Harald Bjørnstad and Airman

Northrop "Nomad" aircraft of the Royal Norwegian Air Force's 'Little Norway' training centre, being raised from Lake Ontario, Toronto, Ontario, 23 June, 1941. PHOTO: LIBRARY AND ARCHIVES CANADA / TORONTO TRANSIT COMMISSION COLLECTION / A054037.

John A. Anderson were flying about a kilometre and a half above Clarkson, a town 30 kilometres from Toronto. Their Douglas attack bomber went out of control, spun in large circles, and came crashing down to earth, narrowly missing a farmhouse. The accident site lay about 275 metres from a major highway, the Queen Elizabeth Way—and dangerously close to traffic there.

A witness recalled:

> I was working in the field when I heard the motor of the plane begin to miss. I looked up and saw it falling and thought it would crash in the field where I was working. As it approached the ground, the motor seemed to pick up again and for a second it looked as if the pilot would pull out of the spin. The motor failed again, however, and it crashed. Immediately I rushed over, but when I arrived the whole machine was enveloped in flames, and the heat was so intense it was impossible to approach closer than fifty feet.[7]

Another devastating plane crash occurred soon afterwards.

July 27, 1941, was a very hot Sunday and thousands were enjoying Toronto's beaches. Kaare Moe was piloting a Fairchild airplane on a routine training flight with two other planes.

The three planes were flying in a "V" formation about one kilometre above the earth. As the aircraft were changing to an echelon formation, the aileron on one wing of Moe's plane jammed, causing his plane's wing tip to touch the wing tip of a second aircraft. Moe's plane started diving towards Lake Ontario. People on the beach first believed the pilot was stunting, and it seemed to them that Moe had brought the plane out of the stunt, but then the aircraft headed nose first into the water.[8]

Oluf Reed-Olsen, Moe's fellow saboteur, and the man who had escaped to England with Moe in a small fishing boat, wrote: "In him I lost a friend, a comrade, and a man whose like I had not met before and have not met since."[9] Kaare Moe was 24 years old.

The concerns expressed by the Coroner's Jury relating to the safety of Toronto's citizens, followed by the subsequent deaths of three Norwegian airmen, made it clear to everyone that the RNAF needed to move camp.

The Norwegians began searching for a new home.

Overseas successes

As Little Norway celebrated its first anniversary in November 1941, the camp received reports of the Norwegian airmen's successes overseas. Speaking to the press, Lieutenant Colonel Ole Reistad said that the air crew who trained at Little Norway, now flying fighter planes with the RAF's Norwegian Squadrons 331 and 332, had sunk two Nazi submarines, blown up a munitions train in France, and shot down several German bombers.[10]

Lieutenant Colonel Reistad explained that the Norwegian flyers in England were specializing in a daylight-raiding technique involving low-flying. By flying only a few feet above the ground, the air crew could direct bombs and gunfire with greater accuracy. "The low-flying technique is hazardous, and some of our boys have been bringing their planes in with incredible damage," said Reistad. "One had three-quarters of his tail surfaces shot away," he noted.[11]

The RNAF also received praise for the Norwegian airmen who had joined the RAF Squadron 330, based in Iceland. This squadron conducted reconnaissance patrols, protected merchant shipping, and escorted convoys in the Battle of the Atlantic. The RAF's Commanding Officer in Iceland applauded the Norwegians for their skilful collaboration, and the accu-

Radio School.

PHOTO: LITTLE NORWAY MEMORIAL.

racy with which they carried out their orders. In fact, the air crews flying from Icelandic bases completed some of the most dangerous missions of all, fighting Arctic winds, blizzards, and ice storms in their small Northrop floatplanes.[12]

At the same time as praise was flowing for the Norwegian squadrons, there was some sad news. Lieutenant John E. Løken of Squadron 331, considered one of the best Norwegian pilots and a veteran of the war in Norway, was lost over the North Sea. His Hawker Hurricane had developed engine trouble. Løken bailed out and was last seen floating in a dinghy, but rescue vessels failed to locate him. While training at Little Norway, Løken had met and married Johanne Klein, daughter of Denmark's consul to Canada.[13]

The year 1941 ended with the United States entering World War II after Japanese aircraft attacked Pearl Harbour, Hawaii, early on December 7, 1941.

More accidents

Crown Prince Olav and Crown Princess Märtha visited Little Norway on January 20, 1942, to open a Wireless Operators' School called "Radio City" for training pilots and navigators.

Nortraship needed additional radio officers for the Norwegian merchant marine ships. Potential radio officers were chosen from among experienced sailors and sent to Little Norway. While there, the sailors wore RNAF uniforms and took basic training. On graduation, the new radio officers returned to the merchant marine.[14]

The Crown Prince and Crown Princess also dedicated a gymnasium known as "Haraldshallen" or "Harald's Hall," named in honour of five-year-old Prince Harald.

That day, a number of Norwegian flyers staged a mock battle and flying demonstration over Lake Ontario to entertain the Royal visitors. A couple of hours later, one of these pilots, Sergeant Peter Christian Ring, was flying with five other Douglas bombers in formation, when his aircraft went into a spin and plunged to the ground, travelling at a speed of over 600 kilometres per hour. The crash occurred at a farm near Oakville. It is believed that Sergeant Ring died instantly.[15]

On January 24, 1942, Lieutenant Trygve Ola Nalki left Toronto Island Airport in a Curtiss Hawk on a training flight. Nalki was a veteran of the war in Norway, where he had battled the Germans at Narvik; he had nearly completed his advanced flight training at Little Norway.

Nalki disappeared that day, along with his aircraft, and neither were ever found. Someone reported seeing a parachute falling into Lake Ontario near Clarkson; others reported observing a plane suddenly dive into the water or woods close to the Scarborough Bluffs. Authorities believed that Nalki's plane developed motor trouble, which left him unable to radio for help or to reach land, and that the plane subsequently crashed into Lake Ontario. A search was called off after 72 hours.[16]

Sergeant Sven Omre was taking private flying lessons at Barker Field, Toronto, while waiting for flight training at Little Norway. However, on his first solo flight, Omre's Cub airplane collided with a Tiger Moth flown by RCAF pilots. Omre died in the subsequent crash.[17]

Lieutenant Thorleif Bergan, age 36, died in an aircraft crash near Streetsville on October 1, 1942.[18]

Close calls continued to be numerous. On February 12, a Norwegian plane carrying two trainees made a forced landing in a field at Bathurst and Fleet Streets, Toronto. The men were unhurt although the aircraft was damaged.

The same day, a woman driving along Lake Shore Boulevard West, near the Toronto waterfront, saw an aircraft make a forced landing at Exhibition Park. Its pilot climbed out of the wreckage and ran to her car. He refused to be taken to the hospital and, instead, asked to be driven to his senior officer at camp. She told a reporter that she was so excited that she didn't even consider where she might have to go, but had it been Camp Borden or Petawawa, she'd have driven him there. She took the pilot to the gates of Little Norway. He left the car, saluted her, and made his way to his senior officer.[19]

Haraldshallen, or Harald's Hall.

PHOTO: LITTLE NORWAY MEMORIAL.

20

Muskoka Airport

During the Depression in the 1930s, the Canadian government embarked on a number of make-work projects to help reduce the rampant unemployment across the country.

On several occasions, the Town of Gravenhurst had approached Canada's Department of Defence, asking that an airport be built at a site about nine kilometres northeast of Gravenhurst. As one of its Depression make-work projects, Canada decided to create an emergency landing field there as part of the Trans-Canada Airway System, as had occurred in Emsdale.[1]

A.E. Jupp Construction Co., creating airfield at Muskoka Airport.

PHOTO: GRAVENHURST ARCHIVES.

The federal government purchased the property from Hart Crawford and others, and work began on the landing field in 1931. The men clearing the property lived in a camp set up by the field. They earned two dollars a month, plus their board.[2] The airfield was completed in June 1936, when the Civil Aviation Division of the Department of National Defence took over daily operations.[3]

The Town of Gravenhurst asked that the airfield be designated "Ferguson Airport" after the first person to privately own an airplane in Muskoka, Arthur Ferguson. However, the federal government's policy was to name airports after the nearest postal outlet and so it was first called "Reay Airport." The town wasn't pleased and requested that the name be changed to "Gravenhurst Airport." After some negotiations, officials from Gravenhurst and Canada reached a compromise and the airfield was christened Muskoka Airport in 1938.

After World War II broke out, the RCAF used Muskoka Airport as an auxiliary base for Canadian Forces Base Borden. Anson, Harvard, and Yale training planes came and went from the airfield during the summer in 1940 and 1941, until the RCAF had the ability to conduct flight training on its own landing fields.[4]

Move to Muskoka

Lieutenant Colonel Ole Reistad had visited Muskoka Airport on one of his trips to Vesle Skaugum, and decided that the airport would be a good place for Little Norway's new home.[5]

The population of Muskoka was about 20,000 at the time, and there was little air traffic, making it a safer spot to train flyers than in Toronto.

Muskoka Airport didn't possess a waterfront to accommodate seaplanes, but this feature no longer mattered. Once the Navy and Army Air Services had united into one air force, a separate Norwegian naval presence had ended in Canada. The final Elementary Flying Training course for naval trainees, given at the Toronto Flying Club, finished in May 1941.[6]

On January 1, 1942, the Norwegians signed a contract with Canada to lease Muskoka Airport for the duration of the war. One month later, the RNAF purchased an adjacent farm, some 131 hectares in size, owned by Thomas Short. The purchase price was $4,000 and title to the property was taken in the name of one of the RNAF officers.[7]

Lieutenant Colonel Reistad told reporters on February 21, 1942, that the major part of the RNAF would be transferred to Muskoka within the next few weeks. One reason for the move, he said, was the constant public criticism of the Norwegians' flights over Toronto, which couldn't be helped with a base located so close to the city. Reistad also mentioned that the lack of hangar space at the Toronto Island Airport meant the Norwegian aircraft had to be parked outside during the coldest weather. Finally, he said, they wished to largely eliminate the possibility of crashes involving young pilots.

Elementary Flying Training School would take place at

Royal Norwegian Air Force hangar under construction, Muskoka Airport.

PHOTO: GRAVENHURST ARCHIVES.

Muskoka Airport where, according to Reistad, "conditions are almost perfect for this kind of training." But Reistad stressed that Little Norway in Toronto wouldn't be abandoned, as Advanced Flying School would continue at Toronto Island Airport. He said the RNAF was very sorry to leave the city, for people had been wonderfully kind, but there was room to expand in Muskoka, whereas every square foot of space was already utilized at Little Norway in Toronto. Reistad continued:

> There are naturally too many restrictions in the Toronto area for a highly successful flying training program. The lads have to abide by low-flying regulations over the city and there is always the danger to the public in the event of a crash. In

Some famous people joined the Royal Norwegian Air Force, including French Prince Jacques de Bourbon-Parma, at right. Here, he speaks with Crown Princess Märtha, and his mother, Denmark's Princess Margrethe.

PHOTO: LITTLE NORWAY MEMORIAL.

> Muskoka those dangers do not exist. It is especially more suitable there for elementary flying training. In addition, the boys at Little Norway have a great many diversions from their work, such as entertainment and social life. At Muskoka they make their own fun and the rugged countryside lends [itself] to it nicely. They can ski to their hearts' content and practice shooting.[8]

No doubt there was an additional reason for the move: discontent with the price of gasoline and oil charged by the Toronto Harbour Commissioners. Some officers told reporters that the savings in gasoline costs that could be reaped by moving to Muskoka would help pay the cost of constructing the new camp.[9]

Famous recruits

Little Norway's trainees included a French prince and a South Seas adventurer.

Prince Jacques de Bourbon-Parma was the son of Danish Princess Margrethe and French Prince René de Bourbon-Parma. After the Germans occupied France, the family escaped to the United States. Though a French citizen, Prince Jacques volunteered for the RNAF and trained as a pilot at Little Norway.[10]

Norwegian adventurer Thor Heyerdahl was living in British Columbia, studying First Nation rock carvings, when the Germans invaded Norway. He travelled to Toronto and joined the RNAF in 1942, training as a radio officer. Heyerdahl, his wife, and two sons often spent time at Vesle Skaugum.

In 1947, Heyerdahl and a mostly Norwegian team would sail a small wooden raft, the *Kon-Tiki*, from Peru to the Cook Islands to support his theory that Peruvians had populated the South Sea Islands.[11]

Kai Pedersen and Pat Halden

Patricia Halden, 22 years old, returned to Toronto on Boxing Day in 1942, after visiting her family in St. Catharines. She was a beautiful girl, who modelled for artists and photographers, and worked as a receptionist at an insurance company. Pat had been dating a member of the RNAF who was supposed to meet her at Union Station, but he wasn't there when Pat's train arrived.

A lot of people had been smoking on the train and Pat, who didn't smoke, was feeling ill from breathing in the second-hand smoke. She seemed to have lost her voice. She decided to telephone the airman to cancel their date. Pat made her way to the Royal York Hotel, where there were telephones in the lobby, but she wondered how she would make herself heard during the call.

Kai Pedersen, from Bodø, was a member of Norway's merchant marine. After the Germans occupied Norway, his ship made dangerous voyages between North America and Europe on behalf of Nortraship.

Kai had been sent to Little Norway to take a three-month-long radio operator's course before returning to sea. He happened to be standing in the Royal York's lobby as Pat entered.

Pat noticed the tall, handsome young man wearing the RNAF *uniform. She thought that this man would likely know her Norwegian date, and decided to ask him to place the telephone call to Little Norway for her. As Pat explained her situation to Kai, her headscarf fell down around her shoulders. Kai later said that this was the moment when he decided that Pat was not going to date anyone else but him.*

Kai called the telephone number Pat had given him and spoke quickly in Norwegian, then told Pat: "He's not there. I'd buy you a drink, but I have no money." Pat said that would be fine, as she didn't drink alcohol anyway.

They sat down in the bar, Pat sipping a ginger ale and Kai drinking a beer. He captivated the young woman with his views about philosophy and the great writers. Before Pat knew it, the bar was closing and they left, walking out into the cold weather together.

Kai pursued Pat over the weeks to follow, which was a somewhat risky move on his part as Pat's former boyfriend was one of Kai's superior officers. Before long, they had decided to marry.

The commanding officer's consent was needed before an airman could marry. RNAF *regulations also required that the airman be at least 21 years of age, have six months' service, and furnish health certificates for both himself and his prospective bride. Trainees couldn't marry until they finished their courses, except in special circumstances.*

Kai Pedersen.

PHOTO: PATRICIA PEDERSEN COLLECTION.

Kai took Pat to meet Lieutenant Colonel Ole Reistad to seek his permission to wed. Pat remembers Reistad as "a wonderful person." He gave his approval.

Pat and Kai were married in the Grace Church on the Hill in the Forest Hill area of Toronto on May 29, 1943.[12]

21

Espionage

During 1942, Germany faced new challenges in the war, including the United States's entry into the conflict and the Soviet Union's successful attacks on the Eastern Front. In addition, Allied commandos had raided the Lofoten Islands again.

With his attention diverted to these fresh threats, Hitler decided he wanted no further trouble in Norway. In an attempt to create calm, Josef Terboven brought Vidkun Quisling back politically, naming him "Minister-President" of Norway and giving him leadership of the Nasjonal Samling party. Terboven kept the real governing power, however.[1]

The Quisling Government imitated Nazi objectives and policies to prove its worth to Hitler, including setting up a youth organization akin to the Hitler Youth, known as the *Nasjonal Samling Ungdomsfylking* or NSUF. Every able-bodied boy and girl, aged 10 to 18, was required to take part in a youth association under the NSUF's umbrella. Immediately, teachers, parents, and the Church of Norway strongly objected.

In addition, the Government established a Nazi teachers' association and ordered all teachers to join. Some 12,000 of Norway's 14,000 teachers protested against both the NSUF and the new teachers' association. Their opposition led to arrests: on March 20, 1942, 1,100 teachers were arrested and taken to detention camps. Another 200 arrests took place in the next couple of weeks.

In April 1942, soldiers forced 500 of the arrested teachers to board a steamer that was only big enough for some 150 people; the boat left Trondheim and sailed to Kirkenes in the country's extreme north. Many of the teachers were already sick or worn out from their time in the detention camps, and faced extraordinarily bad hygienic conditions on the steamer.[2] At Kirkenes, the teachers worked at hard labour, surviving on a starvation diet and living in icy barracks, until allowed to return to their homes in November.

The Lutheran Church also rebelled against the Nasjonal Samling's attempts to make use of it for Nazi purposes. Its seven bishops resigned, saying they would continue their work separately from the state church. All but a small minority of Norway's ministers and priests resigned their positions, too. In April 1942, reportedly only 30 of over 1,000 churches in Norway held Sunday services.[3]

Shortly after Easter, Bishop Eivind Berggrav, the Lutheran Primate, was arrested and held in a concentration camp until released to his cabin in the forest near Oslo, where he spent the rest of the war in solitary confinement.

There were around 1,700 Jews living in Norway when the Germans invaded; arrests of Norwegian Jews began in the fall of 1942. In October, police rounded up Jewish men over 15 years of age in Trondheim and Oslo. During the night of November 25–26, all other Jews in Oslo, including women and children, were arrested. On November 26, over 500 Jewish prisoners were taken to Oslo's harbour, where they were loaded onto the German transport ship *Donau*. The ship landed in Stettin, then trains transported the prisoners to the Auschwitz-Birkenau concentration camp in Poland.

Deportations continued intermittently over the course of the war. In total, around 760 Jews were sent to concentration camps, where all but a handful died. Some Jews went into hiding, and about 900 escaped from Norway to Sweden with the aid of the Resistance.[4]

The protests by parents, teachers, and the clergy against the Nasjonal Samling's initiatives irritated Hitler, and he blamed Quisling. When Quisling tried to contact Hitler to negotiate Norway's future political status, he was advised that Hitler wanted no such discussions until after the war ended. Quisling was also told not to approach Hitler directly anymore, but to contact him through Terboven.

Hitler believed the Allies would invade Norway at any time to gain a foothold from which to launch an attack on Germany, and he ordered a colossal build-up of military forces in the country.[5]

Oluf Reed-Olsen, alias Oscar Leonard Hansen,
operator of radio station, Kristiansand, Norway, 1943.

PHOTO: OLUF REED-OLSEN FAMILY COLLECTION.

Tempsford Taxis

Many Norwegian secret agents, who had been trained in the United Kingdom, arrived in Norway during 1942. They often came by sea, hidden on fishing boats ("The Shetland Bus").

The Tempsford Airfield near Cambridge, England, was the base for many top-secret missions flown by planes that were nicknamed "The Tempsford Taxis." Air crew from Bomber Command, Squadrons 138 and 161, flew Halifax bombers, Westland Lysanders, and other planes; they carried out stealthy missions, dropping ammunition, weapons, radios, and other items to specific locations across Europe, including Norway. The airplanes also dropped off and picked up secret agents.

Operations took place the week before a full moon and the week afterwards, during the darker months of the year, when moonlight assisted the pilots in finding their way. It was very dangerous work: the aircraft had to fly low, and drops had to take place speedily to avoid being detected by the enemy. If

Johan Erland.

PHOTO: SANDRA HOWES COLLECTION.

landing or picking up an agent, the aircraft would usually be on the ground for only a few minutes.

The BBC broadcast coded messages alerting the Resistance to upcoming drops. Once the air crews were over German-occupied countries, they had to rely on visual clues on the ground to find the correct location for the delivery, as any radio contact between the aircraft and the patriots below would have divulged the operation to enemy listeners.

At the allotted time, members of the Resistance would mark the drop-zone with landing lights and the drop would take place. Sometimes the air crew were accidently misled by other lights in the area, and sometimes they were misled on purpose by the enemy, who had learned about the upcoming drop and set up lights of their own.

A number of Norwegians who had trained at Little Norway were assigned to 161 Squadron, including Per Hysing-Dahl, Johan Erland, and Egil Sandberg. Over the course of the war, the RAF lost 126 Tempsford Taxi airplanes, but none of these had Norwegian air crews. When asked why the Norwegians managed to avoid the fate of 126 other aircraft, Egil Sandberg replied: "It was Little Norway that made the difference. The training there was superb."[6]

The mayor of Blois, France, in later years, called the pilots "heroes—flying by the light of the moon, without radio navigation, without information about power lines, or other obstructions—And who would be waiting for them, Gestapo or Resistance forces?"[7]

An agent who landed in Norway customarily worked with local Resistance groups, teaching them espionage and establishing radio contact with the Allies to convey intelligence about German warships. However, the Gestapo knew these espionage activities occurred and hunted for the agents and their Norwegian contacts unceasingly.

Reprisals were brutal. In Telavaag, when the Gestapo attempted to arrest two SOE agents, a gun fight ensued. One of the agents and two Gestapo men were killed, while the other agent was wounded. The Germans executed the wounded agent, along with 18 young Norwegian men. They also destroyed all 300 homes in Telavaag, sunk all the fishing boats, and killed, or took away, all the cattle. Men aged 16 to 65 years old were sent to concentration camps in Germany, where most died, and the remaining populace was interned in Norway.[8]

LEFT Gunnar Sønsteby. PHOTO: NORWAY'S RESISTANCE MUSEUM.
RIGHT Max Manus. PHOTO: UNITED KINGDOM GOVERNMENT.

Max Manus, Gunnar Sønsteby, and Oluf Reed-Olsen

Maximo Guillermo "Max" Manus from Bergen was a member of the Norwegian Army and fought during the war in Norway. After his country surrendered, Manus helped organize the Resistance until he was captured by the Gestapo. After escaping custody, Manus made his way to Sweden, and from there, travelled through the Soviet Union, Turkey, Syria, and other countries to Suez. During the summer of 1941, Manus sailed from Suez to New York City on the EMPRESS OF ASIA, *a ship, which, coincidentally, also carried Egil Chramer and Per Hvidsten.*

Manus spent several months training with small arms at Little Norway in Toronto; he then travelled to the United Kingdom where he learned sabotage, disguise, and parachute skills. He joined the SOE's Norwegian Independent Company No. 1.

After completing his espionage training, Manus and another saboteur parachuted into the countryside outside Oslo. Manus became an expert at sabotaging German warships, sinking and damaging many by the use of homemade limpet mines. These were magnetic mines that attached themselves to ships' hulls. The mines were detonated by time fuses. His most famous successes include an operation in 1945, when he and others set limpet mines on the German transport ship, DONAU, *resulting in its scuttling.*[9]

Gunnar Sønsteby fought in Norway's Resistance movement almost from the first day of the German invasion. He worked in Norway with the British SOE under the code name "No. 24," reporting information about the movement of German armed forces in Norway and arranging for drops of radio operators and transmitters in various towns. Sønsteby had a network of hideouts and agents, and used disguises and cover names. In spite of this, the German authorities pursued him, and by May 1943, he had to escape to England. At SOE training camps there, Sønsteby took commando training. He learned parachuting, use of explosives, and other espionage skills. However, he was keen to return to his work in Norway.

On November 17, 1943, in a Tempsford Taxi flown by Per Hysing-Dahl and his crew, Sønsteby and fellow agent Knut Haugland parachuted into Kongsberg, Norway.[10]

Oluf Reed-Olsen had graduated from Little Norway's naval pilot courses and earned his wings in 1941. He journeyed to Scotland with the first members of Squadron 330 and served in Iceland and then the Shetland Islands. While there, the SOE contacted him and he was sent to Espionage School in England.

After training, Reed was taken to Tempsford. He remembered one particular thing about dinners at the country home there, where the secret agents awaited their missions—something called the "Operational Egg." It was extremely hard to obtain fresh eggs in England at the time. Sometimes, at supper, two fresh eggs would be carefully laid on a dinner plate and served to an individual. That meant that it was the individual's turn to take part in the night's operations: the eggs were a last little extravagance for those who might never come home. Reed wrote that no matter how much a person liked fresh eggs, his or her appetite would seem to vanish at the sight.

On April 17, 1943, Reed found two fried eggs on his plate and was soon in a Halifax airplane, heading for a parachute drop into the forest near Drammen, Norway. Due to weather conditions, the drop didn't take place for a couple of days, and, when it did, Reed landed at the top of a snowy tree with a dislocated knee. In extreme pain, he used a stick to slip his knee back into position and crawled down to earth.[11]

22

Dieppe

Squadron 330, operating out of Iceland, often faced extreme weather conditions in the Arctic region, and a number of their Northrops were lost at sea.

In April 1942, the squadron's commanding officer, Captain Hans A. Bugge, and his crew were conducting a sweep for U-boats in the waters off Iceland when their Northrop disappeared. Neither Bugge nor the crew were ever found. Just a month earlier, Captain Bugge had received Britain's Air Force Cross for acts of valour during active operations.[1]

The RAF provided Squadron 330 with Consolidated Catalina seaplanes during the summer of 1942. The twin-engine Catalinas offered big improvements over the single-engine Northrops, including longer flight range, better armoury, and increased endurance.

But as the Americans became more engaged in the Battle of the Atlantic, they began to build bases in Iceland. Soon there was no need for the RAF to maintain airfields there. In January 1943, the RAF transferred Squadron 330 from Iceland to Oban, Scotland, and then to Sullom Voe in the Shetland Islands, where it would remain for the rest of the war.

In Scotland, the squadron flew powerful, four-engine Short Sunderland flying boats, escorting Allied convoys to spots in the far north, conducting rescue missions, and patrolling the ocean for enemy ships and submarines.

The RAF established a new Norwegian unit, Squadron 333, on May 10, 1943. This squadron possessed three Catalinas; the RAF lent it other aircraft, including six De Haviland Mosquito bombers and some training planes. The squadron was composed of two divisions: A Flight, flying Catalinas and based in Woodhaven, Scotland; and B Flight, operating Mosquito bombers and based in Leuchars, Scotland.

While both flights carried out patrols, attacked German ships and planes, hunted for U-boats, and conducted search and rescue missions, A Flight flew secret missions, conveying and retrieving secret agents, and delivering supplies and radios to the agents.

Rolf Hauge

When Rolf Hauge had arrived at Little Norway, he told the authorities that he wanted to get back to the war as soon as possible. Most recruits wanted to fly planes, and with more and more Norwegians arriving regularly at Little Norway, there was a long list of people already awaiting pilot training. Rolf had studied mechanics while at school in Norway, so he asked the RNAF to allow him to take classes to become a flight engineer and air gunner. After less than a year of training in Canada, he travelled to Scotland and joined Squadron 333, A Flight.

Rolf remembered that a few days before Christmas 1943, a

ABOVE Rolf Hauge of Norwegian No. 333 Squadron taking gunnery training. PHOTO: LITTLE NORWAY MEMORIAL.

LEFT Rolf Kolling, pilot with Norwegian No. 332 Squadron. PHOTO: ROLF KOLLING COLLECTION.

number of Catalinas left the United Kingdom and flew over the North Sea to Norway's islands, keeping very low to avoid detection by German radar. Rolf's plane flew so low it seemed to skip over the waves. No Germans were expected to be on the islands so the journey was relatively safe for the air crew.

The roar of the Catalinas' engines brought out the local citizens, and the air crew waved a big Norwegian flag. The squadron dropped bags of chocolates, cigarettes, coffee, and other treats to the people below. The air crew had a great time, and managed to head back home before the enemy discovered their arrival.[2]

Squadron 331 and 332

In 1942, Captain Helge Mehre led Squadron 331; Major Odd Bull commanded Squadron 332 until April, when Wilhelm Mohr took over. By July 1942, both fighter squadrons were operating from North Weald, England, as part of Group No. 11, RAF.

The RNAF originally planned to bring Little Norway's Curtiss and Douglas aircraft overseas for use by Squadrons 331 and 332, as had occurred with the Northrops flown by Squadron 330. However, the amount of technical support needed to keep the fighter planes operating was deemed excessive, given the relatively small number of aircraft involved. Instead, the Norwegian air crew took conversion flight training, learning to fly Hawker Hurricanes and then Supermarine Spitfire fighters.[3]

At first, the squadrons received older model Spitfires, inferior to German Focke-Wulf and similar aircraft. However, before long, newer versions of the Spitfire arrived at North Weald.

Rolf Kolling

Rolf Kolling joined Squadron 332 in late 1942. Before departing from Canada, as he had leave time, Rolf decided to join some air force friends who were going to Montreal for a farewell party. Just after taking off from Toronto Island Airport in a Curtiss Hawk, Rolf felt that something was wrong with the plane's engine—it seemed to be stalling. He flew back to the airport and landed just before the engine stopped. Rolf's pilot logbook shows "eight minutes flying." This was Rolf's last flight in Canada and he felt very lucky to have avoided a crash; however, he missed a great party in Montreal.

Once in North Weald, Rolf began flying missions over occupied Europe and often encountered enemy aircraft. He recalled how the Luftwaffe fighters customarily flew very high and swooped down to attack the RAF planes below them. But when Squadrons 331 and 332 received the new Spitfire Mk IX aircraft, circumstances changed for the better. These planes possessed more powerful engines, could move faster, and could climb to over 12,000 metres. With the Spitfire Mk IX, the RAF could more than meet the challenge of the Luftwaffe's fighters.[4]

Norwegian No. 331 Squadron in November 1942. Top row: Johannes Greiner, Martin Gran. Second row, from left: Helge Sognnæs, Leif Lundsten, Stein Sem, Knut Backe, Anton C. Hagerup, Rolf Arne Berg, Philip Yateman, Rolf Engelsen, and Svein Heglund. Bottom row: Reidar Haave Olsen, Kristian Nyerrød, Fredrik Fearnley with dog Varg, Kaj Birksted, and Tarald Weisteen. 1942. PHOTOGRAPHER: JON TVEDTE, *FLYVNINGEN. DET MODERNE EVENTYR*, OSLO, 1958.

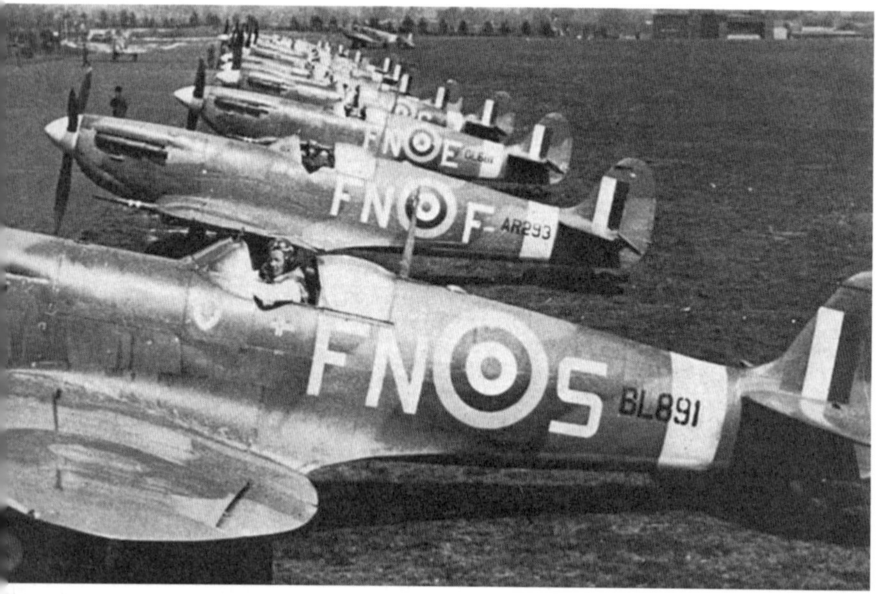

Spitfires

People sometimes said that a pilot put on a Spitfire like a jacket—and that the pilot became part of the plane.[5] Many of the most senior pilots in 331 and 332 Squadrons had a certain Spitfire they called their own. Sometimes they painted a girlfriend's name on the aircraft, or kept score of the number of enemy planes shot down.

Squadron 332's first fatality occurred in June 1942. Sergeant Per Schonberg had bailed out of his Spitfire after it developed engine problems following a dogfight with a German Junkers JU88. He was never located. Sergeant Schonberg had been engaged to marry a Toronto girl.[6]

Other casualties included Captain Jens S. Hertzberg, who had been an organizer and chief of staff at Little Norway. He was 32 years old, which was older than most of the RNAF pilots. After repeated requests, however, he had been allowed to join 332 Squadron and take part in air raids. His Hurricane aircraft was reported missing after it experienced engine trouble returning from a heavy dogfight. It's believed Hertzberg's plane went down over the English Channel.[7]

TOP Spitfire Mk Vs of Norwegian No. 331 Squadron at Catterick, England, on their way to North Weald, England, spring 1942. PHOTOGRAPHER: JON TVEDTE, *FLYVNINGEN. DET MODERNE EVENTYR*, OSLO, 1958.

BOTTOM Some members of Norwegian No. 332 Squadron at Catterick, England, April 1942. Standing, from left, Erik Haabjørn, Johan Kristoffer Bjørn Ræder. Seated from left, Werner Christie, Peder Mollestad, Jan Egil Løfsgaard, Jan Christie, Per Schønberg, and Marius Eriksen. PHOTOGRAPHER: OLE FRIELE BACKER, *KRIGSFOTOGRAFEN*, BERGEN, 1946.

Dieppe

On August 19, 1942, the Allies launched an attack against the German-occupied port of Dieppe, France. The attack was also known as "Operation Jubilee." Its goal was to take control of the city for a short time, secure its perimeter, destroy the port defences, and withdraw by sea. Some 5,000 Canadian troops formed the greater part of the landing force, accompanied by approximately 1,000 British troops and soldiers from other Allied nations.

Dieppe was a disaster. Many things went wrong, such as the late arrival of tanks, and the Allies' failure to send sufficient air and naval bombardment to support the men on the beaches. The enemy fought fiercely. As Squadron 331 pilot Svein Heglund recalled: "The Germans at that time were at their very best."[8] Almost 60 per cent of those who landed at Dieppe were killed, wounded, or captured, and the entire raid was over by midday.

The Luftwaffe appeared above Dieppe in large formations, attacking the Allied naval ships below. The Spitfires soon engaged in numerous dogfights with German Focke-Wulfs and Messerschmidts. Squadrons 331 and 332 took part in the Dieppe raid and distinguished themselves.

Due to limitations on their fuel capacity, the Spitfires had to return frequently to North Weald and refuel before returning to the battle zone. Once back at home base, ground crews moved with lightning speed to rearm and refuel the planes so they could take to the air again. When Wilhelm Mohr returned from the first sortie, he tried to conceal the fact that a bullet or piece of shrapnel had gone right through his leg. Others noticed that Mohr's boot was full of blood. Mohr wanted to continue flying after his leg had been treated, but

Three pilots from the RAF's 331 and 332 Norwegian Squadrons were decorated with Norway's War Cross with Sword following the Dieppe Raid. Left to right: Helge Mehre, Crown Princess Märtha, King Haakon, Wilhelm Mohr, Crown Prince Olav, and Kaj Birksted. These three pilots also won Britain's Distinguished Flying Cross. PHOTOGRAPHER: OLE FRIELE BACKER, *KRIGSFOTOGRAFEN*, BERGEN, 1946.

Group No. 11's wing commander refused to allow him to return to the air. Captain Finn Thorsager took over command of Squadron 332 for the rest of the day.[9]

During the day, the squadrons flew four sorties over Dieppe, rotating pilots so that each one flew two missions. The last sortie defended the retreating convoys of troops.

The RAF squadrons kept tally of damage to enemy aircraft according to the number of aircraft they destroyed, the number they probably destroyed, and the number they damaged. Squadron 331 claimed six German aircraft destroyed, four probably destroyed, and six damaged. Two of the Norwegians' Spitfires were lost, but their pilots were safe. Squadron 332 claimed eight Luftwaffe aircraft destroyed, one probably destroyed, and five damaged. Five of Squadron 332's Spitfires were lost. Three of the pilots of these destroyed Spitfires were safe and two had been captured. Of the pilots captured, Sergeant Per Bergsland became a prisoner of war at Germany's Stalag III. In 1944, Bergsland and a pilot from Squadron 331, Jens Müller, took part in "The Great Escape" from that camp and successfully made their way to safety.[10]

The Norwegians had done an excellent job; their total scores represented about 15 per cent of all enemy planes destroyed in the Dieppe air battle.[11] A British RAF commander wrote to Hjalmar Riiser-Larsen, the RNAF's commander in chief, saying, "You can with justice be proud of your boys."[12]

Three RNAF pilots received the British Distinguished Flying Cross for their actions over Dieppe: Wilhelm Mohr, leader of the 332 Squadron; Helge Mehre, leader of the 331 Squadron; and Kaj Birksted, second in command in the 331 Squadron.[13]

In 1943, the RAF named Squadron 331 its top-scoring Allied Fighter Squadron. Squadron 332 had the third highest score.

Norwegian Air Force serving in other roles

The RNAF contributed to the Allies' war effort in a number of other roles, too, such as flying with the RAF's Fighter Command, Ferry Command, and Bomber Command.

Ferry Command was based at Dorval Airport, Montreal. From there, air crew flew heavier aircraft like Lockheed Hudsons, Boston Bombers, Consolidated B-24 Liberators, Lancasters, and others to Europe. The airplanes usually made stops at Gander or Goose Bay in Newfoundland, then at Greenland and Iceland before landing in Prestwick, England.

By late July 1942, pilots trained at Little Norway had ferried over 25 planes across the Atlantic Ocean.[14]

That year, Lieutenant Colonel Ole Reistad travelled to Britain to visit the RNAF training bases and meet with government officials. While at one of the Ferry Command bases, Reistad watched two bombers arrive from America. When the pilots disembarked, and Reistad discovered they were graduates of Little Norway, he remarked, "It was wonderful to see our boys bring over those huge machines."

Regarding Squadrons 331 and 332, Reistad marvelled that the Norwegian airmen, who only a few months earlier had been flying in slow, unspectacular circles over Toronto harbour, were now in hot pursuit of German aircraft over France and Holland. The Norwegian flyers were being used extensively to attack enemy shipping in the English Channel and took "great delight in swooping down on railway trains along the coast," he said.[15]

John Stene and Per Waaler

In 1942, John Stene and Per Waaler were transferred from Little Norway to Ferry Command in Montreal, where they learned to fly Bostons and other heavy twin-engine aircraft. John later said he preferred flying a Boston to any of the other 20 kinds of aircraft he flew during the war. However, it had one serious flaw: the radio and intercom were very poor and usually quit. When that occurred, a pilot wishing to communicate with a navigator had to write a mes-

sage on paper, clip it with a clothes peg to a wire running between two pulleys, and then tug on the wire to roll the message along to its recipient.

Weather on the northerly routes could cause extremely long delays. In February 1943, John Stene had to wait 32 days in Greenland for the weather to improve enough to fly on to England, resulting in a trip that lasted a total of 36 days.

Following their service with Ferry Command, John Stene and Per Waaler were transferred to RAF *Bomber Command Squadron 76, based at Linton-on-Ouse, England. In Bomber Command, a new pilot flew two trips with an experienced pilot before flying on his own. John and Per were each assigned to fly with senior pilots on a night flight over Dortmund, Germany. For John, it was a routine trip, with normal flak and searchlights. But while John returned safely, Per and his aircraft did not.*

After about a month, the Red Cross advised that Per was a prisoner of war in Germany, and otherwise in good health. John later learned that a German night fighter had shot down Per's plane. Four crew members had died and four, including Per, had survived.[16]

K
123
153

23

Little Norway in Muskoka

After the RNAF had leased Muskoka Airport and purchased an adjacent farm, the time had come to build a new training camp. Second Lieutenant John Engh, the architect who had constructed the central lodge and other buildings at Vesle Skaugum, returned to Muskoka to design Little Norway.

Due to ongoing funding constraints for the Royal Norwegian Air Force, Engh tried to stretch every dollar. He decided to keep a two-year-old barn, already located on the property, and remodel it to create barracks for 80 men.[1]

Engh added new windows and balconies to the barn, along with systems for air conditioning for summer and heat in winter. The barn's interior featured hardwood floors, which were cheaper to purchase than linoleum at the time. The men, who slept in double-decker bunks, kept their clothes in sizeable individual lockers. Showers and a large sauna were situated on the floor below. Sergeant Svend Clement recalled the winter morning ritual of sitting in the hot sauna and then jumping out into a snowbank.[2]

A *Toronto Daily Star* reporter called the barn a "modern, comfortable home for the airmen."[3] A *Globe and Mail* writer found it "an amazingly fine building ... and it is impossible to believe that it formerly served an agricultural purpose."[4]

Engh also designed a beautiful, two-storey building shaped like a huge "H." Whole logs, taken from trees growing on the site, were used for the structure, and most of the log-joining

ABOVE A new log mess building being constructed at Little Norway.

PHOTO: NORWEGIAN ARMED FORCES MUSEUM, OSLO, NORWAY

OPPOSITE Fairchild training aircraft on runway.

PHOTO: GRAVENHURST ARCHIVES.

work was accomplished by Finnish civilians or members of the RNAF. One wing of the "H" housed a library, a canteen, and the airmen's mess, which was a vast, light-filled room decorated with Norway's coat of arms and photographs of the Royal Family. A Norwegian living in South America donated woven curtains for the windows in the mess. In the middle bar of the "H" lay a bright, clean kitchen. The other wing contained the officers' and sergeants' messes. Engh also created sleeping quarters on the building's upper level.[5]

Other Little Norway buildings included a gymnasium, Link Trainer building, guard house, ground school, hangar, offices, garages, and workshops. The old farmhouse on the site became a 12-bed hospital, infirmary, and dental clinic.

There was a large swimming pool which could also serve as a water reservoir in the case of fire. However, most of the Norwegians preferred to swim in nearby Reay Lake. In a back field, a rifle range was set up.[6]

Lieutenant Hans K. Lund served as Little Norway's commanding officer.

A reporter was struck by the lack of waste at Little Norway: in the clothes depots, new uniforms, shoes, and boots lay ready for use. However, in adjoining rooms, tailors mended worn uniforms and cobblers patched and re-soled footwear. Gifts of woollen sweaters and mittens were carefully stored away until the recruits went overseas.[7]

An easy camaraderie existed among the officers and the airmen. On one occasion when Lieutenant Colonel Reistad made an unexpected visit to the camp, Lieutenant Lund

LEFT Aerial view of Little Norway at Muskoka Airport, looking southeast. The camp was situated among Muskoka's forests and farm lands, avoiding the dangers of flying over populous Toronto. Beaver Creek meanders through the fields in the background. PHOTO: NORWEGIAN ARMED FORCES MUSEUM, OSLO, NORWAY.

FACING PAGE At far right, a reconstructed barn provides barracks for airmen. Beside it is the camp's swimming pool. In the centre of the photo lies an H-style, log structure. Its left arm contains officers and NCOs' messes; its right arm holds the airmen's mess, a canteen and a library. Sleeping accommodation is included on the building's upper level. A kitchen is located in the small building between the two arms.

The building below the H-style structure contains the hospital, dental office, and infirmary. The aircraft hangar can be seen at the top of the photo. Other buildings include a gymnasium, guard house, ground school, Link Trainer building, workshops, garages, and offices. PHOTO: KAREN GREAVETTE COLLECTION.

couldn't locate him. Reistad was finally found sitting at a long table in the airmen's mess, eating and talking with a number of enlisted men. He had wanted no special treatment.[8]

Opening ceremonies

On a chilly May 4, 1942, Little Norway officially opened at Muskoka Airport.

Crown Prince Olav and Crown Princess Märtha travelled to Gravenhurst that morning, accompanied by RNAF personnel in seven motor cars. At the south end of the town, the Royal party was met by Ontario Provincial Police, who escorted them to the town hall for a civic reception. Flags decorated Muskoka Road, Gravenhurst's main street, and a crowd of over 2,000 people, including schoolchildren, Boy Scouts, and Girl Guides, gave the Crown Prince and Crown Princess a rousing welcome.

The Union Jack and Norwegian flag flew from the town hall's flagpole; the Gravenhurst Citizens' Band played patriotic songs. In at least one restaurant, Norwegian flags decorated the window and a sign read, "Velkommen Norwegians," or "Welcome Norwegians."

From a stand set up in front of the town hall, Mayor John Gibson greeted the Royal pair.[9] Crown Prince Olav voiced his

TOP Little Norway's official opening, May 4, 1942. Gravenhurst's Mayor John Gibson escorts Crown Princess Märtha and Crown Prince Olav to the ceremonies, while Gravenhurst's Girl Guides form an honour guard. The Girl Guide, second from the right, is Shirley (Tapp) Barlow. PHOTO: GRAVENHURST ARCHIVES.

BOTTOM Opening ceremonies, Little Norway, May 4, 1942. PHOTO: S.J. REGINALD SAUNDERS, *LITTLE NORWAY IN PICTURES*.

gratitude in response and received an ovation from the crowd. At the end of the ceremony, a small Gravenhurst girl, Nancy Fryer, presented Crown Princess Märtha with a huge bouquet of red roses, and the Princess thanked her with a kiss.

After the band played "O Canada," the Royal party proceeded to Little Norway amid cheers from well-wishers.

When the Crown Prince and Crown Princess arrived at the camp's entrance, they found an arched gateway decorated with evergreens and a banner saying "Welcome to Little Bergen." A small group of men from Bergen had lobbied for the camp to be named after their home city and had stealthily hung the banner shortly before the guests arrived at the camp. Their promotion of the name "Little Bergen" met with some success: newspaper articles describing the opening ceremonies published in the *Toronto Daily Star*, *Evening Telegram* and *Globe and Mail* referred to the camp as such.[10]

Crown Prince Olav laughed when he saw the banner but Lieutenant Colonel Ole Reistad wasn't pleased—and the name remained Little Norway.[11]

The Royal couple had lunch in the airmen's mess with a number of military and civilian dignitaries, including Gravenhurst's Mayor Gibson and Bracebridge's Mayor E.A. Whitten. A visiting reporter wrote: "A most excellent but simple luncheon was served, delicious Norwegian koldt bord with cold meats and vegetable salads, followed by perfectly grilled chops and wonderful dessert." The reporter's dinner companion, the camp doctor, told her that many of the cooks came from ships noted for fine food—in fact, a great many had been on the whaling boats.[12]

Per Kaare Nilsen from Oslo was one of these talented cooks. He had worked as a steward in the Norwegian merchant marine. After the German invasion, Nilsen came to

Kitchen, Little Norway.

PHOTO: LITTLE NORWAY MEMORIAL.

Canada and joined the RNAF. He assisted with the planning and building of kitchens and dining facilities at both Vesle Skaugum and Little Norway.[13]

The camp kept pigs for consumption of refuse and for meat, and at one time there were as many as 60 of them. Cooks made use of a smoking oven for pork products such as bacon,

wieners, and sausages. They also smoked salmon. During the summer, volunteers planted a vegetable garden and in the fall, some 250 sacks of potatoes were harvested, along with other vegetables.

Little Norway's kitchens became famous among American and Canadian visitors who rapidly developed a taste for Norwegian dishes. When the camp was busy, the kitchens would serve up to 2,600 meals each day.[14]

After lunch, the opening ceremonies began before a crowd of about 200, many of them Muskoka residents. Gravenhurst's Shirley Barlow was there as part of a Girl Guide troop. She recalled feeling very excited at the opportunity to meet a real prince and princess. Crown Princess Märtha was smartly dressed, Shirley remembered, and the Royal couple was gracious.[15]

When it was time for speeches, Crown Prince Olav thanked Canadian authorities for their assistance in setting up the Norwegian air-training camp for his young countrymen. Air Vice-Marshal George O. Johnston, the RCAF's deputy chief of staff, spoke next. He said, "We have admired your courage; we are proud to call you our friends."[16]

As the RNAF and RCAF flags rose skyward on two flagpoles in front of the main building, Fairchild airplanes thundered overhead, and the camp officially opened.

TOP Barn barracks, Little Norway.

PHOTO: KAREN GREAVETTE COLLECTION.

BOTTOM Sleeping quarters, barn barracks, Little Norway.

PHOTO: LITTLE NORWAY MEMORIAL.

RCAF takes over Toronto's Little Norway

By mid-1942, German patrols had made it increasingly difficult for young Norwegians to escape from their homeland, and the reprisals visited on the families of those who did get away served as a serious deterrent to potential refugees. As a result, fewer recruits came to Little Norway and this dwindling number meant that the RNAF no longer needed its buildings at the foot of Bathurst Street, Toronto. In April 1943, the buildings were transferred to Canada's Department of National Defence to serve as a manning depot. However, the RNAF continued to use Toronto Island Airport for Operational Training, and its administrative offices remained in the city.

The Toronto Harbour Commissioners wrote a letter to Lieutenant Colonel Reistad, expressing "sincere regret that you find it necessary to move from Camp Little Norway and the earnest wish that our friendly relations may continue."[17]

The Lieutenant Colonel replied:

> It is indeed rather sad to leave a place that has meant so much to us as Little Norway and the Island Airport. There has never in our history been such as place which has so highly increased the friendship between the Canadian and Norwegian people, not only for the time being but also for the future. I personally have never had such a certain feeling of leaving a place in body but not in mind.[18]

Still, difficulties dogged the relationship between the Harbour Commissioners and the RNAF right up to its end. After two fires occurred in aircraft stored in the Toronto Island Airport hangar, there was some finger-pointing. The Airport Manager suggested that the fires might have been caused by RNAF staff failing to turn off switches in parked aircraft, or testing aircraft radios while in the hangar. The RNAF replied that their own investigation concluded the fire resulted from defective wiring in the aircrafts' radios, but then noted that over the past year, there had been several reports of mechanics, from both the Harbour Commission and RCAF, using the radios in the Norwegian aircraft to listen to programs broadcast from local radio stations. In response, the Airport Manager made inquiries among his staff, and they denied any knowledge of such events.[19]

Guard and dog, Little Norway.

PHOTO: GRAVENHURST ARCHIVES.

When the RNAF gave notice one day that it intended to vacate part of the office administration building a couple of days later, the Harbour Commissioners refused to accept the

These Fairchild aircraft from Little Norway are equipped with skis, allowing landings on snow-covered ground or frozen lakes.

PHOTO: NORWEGIAN ARMED FORCES MUSEUM, OSLO, NORWAY.

short notice. They relied on their legal right to receive a full month's notice, and the rent that accompanied it. The General Manager complained to the RNAF, saying a Norwegian officer "spoke rather harshly of our attitude in this matter."[20]

In any event, the RCAF took control of the Norwegian camp in an official ceremony on April 5, 1943. Both Lieutenant Colonel Reistad and the RCAF's Air Commodore, Frank S. McGill, expressed regret that the Toronto facility had proven inadequate.

Reistad said that as a result of the RNAF's labours at Little Norway in Toronto, they had an air force many times more modern and effective than before. He added his thanks to Toronto for its "marvelous hospitality and friendliness."[21]

Some Toronto citizens expressed disappointment at the news that the camp would be moving to Muskoka. One reporter noted that people had come to regard Little Norway as almost part of their own war effort.[22]

In 1944, the federal government turned over buildings at Little Norway to the City of Toronto for use as emergency housing.[23]

24

Bears, Bridges and Northern Romance

Some remarkable animals made their home at Vesle Skaugum and Little Norway.

There were guard dogs and huskies. The latter had been brought from the Hudson's Bay Company in Churchill, Manitoba. The RNAF bred them to pull sleds driven by the Norwegian trainees who were preparing for battle in snowy conditions. Some dogs were trained to haul sleds carrying medical supplies and to tow sled ambulances; others to pull sledges laden with ammunition. The dog-teams could travel up to 80 kilometres a day. At Vesle Skaugum, each dog had its own dog-house and knew its name when called.[1]

A countryman living in the United States gave the camp an elkhound known as "Varg," brother to a dog belonging to former American president Herbert Hoover. Varg became a camp mascot.

But, undoubtedly, the most fascinating animals dwelling at the Norwegian camps were the bears. The RNAF raised at least three orphaned bears over the war years.[2] A black bear cub named "Funny," found in the woods close to the camp, became the most popular and famous mascot. Funny took part in wrestling matches with the men, including with flight mechanic John Baadsvik, who wrestled with the bear regularly in the evening.[3]

The Norwegians were very fond of Funny and the bear appears to have behaved gently with them, often rearing up on his hind feet and placing his paws on the chests of familiar airmen.

Bear and airman wrestling, Little Norway.

PHOTO: KAREN GREAVETTE COLLECTION.

Funny enjoyed drinking beer, preferring Canadian beer to other kinds, and was very fond of Coca-Cola too. When famous Norwegian skater and film star Sonja Henie came to

TOP Funny the bear, Little Norway.
PHOTO: LITTLE NORWAY MEMORIAL.

BOTTOM Fairchilds soar above Muskoka Lakes.
PHOTO: KAREN GREAVETTE COLLECTION.

visit the camp in March 1944, she met the bear. Clapping and laughing, Henie watched as Funny sat on his haunches and drank four bottles of pop, burped loudly, and then stretched out his paws for more.[4]

One weekend, the bear broke into the camp hospital and airmen found him going through a medicine cabinet, hauling out bottles, searching for beer or pop. On another occasion, Lieutenant Harald Olafson returned to his quarters to find Funny sleeping in his bed: the bear had been left there by mischievous recruits.[5]

After the war ended, Lieutenant Colonel Reistad brought Little Norway's two teams of huskies and three bears to new homes in Bardufoss, Norway.[6]

Douglas Johnston

Douglas Johnston was a teenaged student living in Port Carling, a village located some 35 kilometres from Little Norway. He recalled that, except for the odd floatplane, aircraft were rarely seen flying above the village, and so the arrival of the RNAF brought excitement.

Doug remembered times when Norwegian pilots buzzed the Port Carling bridge. He said: "The one or two crew aboard could be plainly seen, just as if they were ready to shake hands. Then off they went with a quick wave and thundering engines into the blue yonder. It was not uncommon for them to repeat the run, much to the shock and fright of the people on the ground."[7]

During the summers, Doug worked as a caddy at a nearby golf course. He recalled once walking home along a fairway with some other caddies, when a Norwegian plane circled above and dropped a small parachute. The boys raced to its landing spot and found a handkerchief secured to a bag of grapes, with a note, "Wave if it's OK to land." The boys waved furiously as they wanted to see the plane land, but the pilot thought better of the idea and went roaring off into the blue.[8]

Residents of Muskoka became quite used to the thunder of the Fairchilds and other aircraft as they flew low overhead and practised flying in formation.

Northern romance

While Lieutenant Colonel Reistad might have believed moving Little Norway to Muskoka would remove the trainees from the city's temptations, he was not altogether correct. Girlfriends from Toronto arrived every weekend to visit the young men, usually travelling to Gravenhurst on the "Passion Train," which arrived about 9 p.m. on Friday night. Per Conradi Hansen, who was stationed at Little Norway, wrote that the girls often travelled to Gravenhurst in pairs, in order to assure anxious parents that one girl would take care of the other. As there were many single young men at the camp, it was easy enough to find dates for any of these travel-companions who were unattached.

After meeting up at the train station, the couples would head for Sloan's or another restaurant on the main street for coffee or pop. When it got late, the girls made their way to lodgings in hotels, summer resorts, or friends' homes, while the airmen returned to Little Norway. Sometimes the airmen rented cottages at Gull Lake for the weekend, where parties continued until the early hours.[9]

The Norwegians also dated local women, often meeting them at spots like Vincent's Restaurant, owned by Rosa Vincent and her son, Cyril. The restaurant was located at the corner of Hotchkiss Street and Muskoka Road in Gravenhurst, near the family's department store.

Date night, Vincent's Restaurant, Gravenhurst.

PHOTO: KAREN GREAVETTE COLLECTION.

On Thursday and Friday nights, people gathered in the restaurant's dining room to dance to music played on a Wurlitzer jukebox, for five cents a song. Rosa had two rules: everyone had to dance in their sock feet to save the floors, and the dance had to end at midnight.[10]

Gravenhurst's Yvonne Jones, who was a young girl at the time, remembered that the music from the dances at Vincent's Restaurant could be heard up and down the main street.[11] The restaurant was so busy that it closed all day Wednesday during wartime, to allow its employees a rest.[12]

Other popular places for evening entertainment included

New Year's Eve, 1943, at the NCO mess, Little Norway.

PHOTO: LITTLE NORWAY MEMORIAL.

Gravenhurst's Muskoka Beach Inn, Port Carling's 21 Club, and Bala's Dunn's Pavilion.

Ron Stoddart from Toronto visited his grandmother's cottage near Bala during the war years. He recalled often seeing members of the RNAF at Dunn's Pavilion, enjoying Big Band music. Ron remembered the airmen as a handsome, polite group of men, who made a terrific impression. He said: "The Norwegians stood out as people you'd like to know." And the Norwegians certainly attracted the attention of many lovely local girls. In fact, after word spread that the Norwegians frequented Dunn's Pavilion, Bala filled up with young women on dance nights.[13]

Many of the romantic relationships were temporary, lasting only until the recruits left Little Norway for training courses elsewhere or were posted overseas. Sometimes, the trainees couldn't bring themselves to tell their girlfriends they were about to depart. As Per Conradi Hansen observed, too often girls sat at Sloan's restaurant, "bewildered and heart broken ... crying and wondering 'Why didn't he tell me?'"[14]

But romance turned into marriage in a number of cases, and wedding bells rang in Muskoka towns. Among the happy Gravenhurst couples were Svend Clement and Rae Greavette, Odd Toldnes and Betty Kenny, Johan Erland and Elsa Strom, John Baadsvik and Evelyn Stark. In Huntsville, Per Hvidsten and Leila Todd married.

Given the number of marriages between the Norwegian airmen and Canadian women, it isn't surprising that housing became scarce in Gravenhurst and Bracebridge. Apartments were hard to find, and, in some cases, families shared accommodation.[15]

John Baadsvik and Evelyn Stark.

PHOTO: BETTY HONNOR COLLECTION, LARRY BAADSVIK COLLECTION.

Barbara Taylor and Egil Chramer

Barbara Taylor encountered Egil Chramer, a dental technician at Little Norway, by chance one day in Toronto. She had attended a meeting and was waiting by the door, when two Norwegians arrived in a nearby elevator. One was Egil, who walked over to Barbara and chatted with her for some time.

Egil called Barbara after Little Norway had moved to Gravenhurst and they began dating. Barbara came to Gravenhurst when there was a party or special event at Little Norway, staying with friends who had a home in the town. When he was on leave, Egil visited Barbara in Toronto.

Barbara remembered swimming in the cool lakes at Vesle Skaugum on summer days and dancing with Egil at the camp in the evening. People jived, boogied, and waltzed to records, under the watchful eye of senior officers serving as chaperones.[16]

TOP A toast. Barbara Taylor and Egil Chramer. PHOTO: BARBARA CHRAMER COLLECTION.

BOTTOM Wedding of Johan Erland and Elsa Strom at the Bracebridge Legion. PHOTO: SANDRA HOWES COLLECTION.

25

Patriotism

After World War II began, British leaders became concerned that Germany would invade the United Kingdom. To ensure there were not a lot of enemy prisoners of war inside Britain in the event of an invasion, Britain needed offshore locations for its prison camps. In 1940, Canada agreed to confine the prisoners of war on Canadian soil until the war ended.

Some 10 prisoner-of-war camps were set up in Ontario, including one at the Calydor Sanatorium, a former tuberculosis sanatorium in Gravenhurst. The new prisoner-of-war camp, named "Camp Calydor" or "Internment Camp 20," was officially designated as an officers' camp, but as Cecil Porter notes in *The Gilded Cage: Gravenhurst German Prisoner-of-War Camp 20*, a sizeable number of prisoners came from other ranks, most of whom were sent to serve the officers as orderlies.[1]

The first prisoners arrived by a special train on June 30, 1940. They hailed from the German Army, Navy, and Air Force and marched resolutely from the train station to Camp 20, as local citizens lined the streets to watch. Hugh Clairmont, newspaperman and musician, remembered the public reaction that day: "There really was a hatred for them at first. People yelled at them—it was just like a chant—they became so emotional. But it wasn't long before people took the camp for granted."[2]

Even so, having an enemy prisoner-of-war camp in town, while local young men were away fighting in Europe, caused an undercurrent of tension. On the steamboat *Sagamo*, the band would play "There'll Always Be an England" and "Rule Britannia" each day as the vessel sailed past Camp 20.[3]

Lieutenant Colonel Ole Reistad told a reporter that the presence of the prisoner-of-war camp in Gravenhurst boosted the morale of the RNAF, as every day they could see their enemy behind barbed-wire fences. The Germans had boasted that the Luftwaffe had wiped out the Norwegian Air Force in one day, Reistad said, but now the Nazi airmen-prisoners could see, with their own eyes, Norwegian planes flying over their prison camp—and know full well where the airmen were heading.[4]

Undoubtedly, the presence of the prisoners of war aggravated the Norwegians. A story lingers of a group of RNAF flyers who had observed that some German prisoners marched along a given street at the same time every day as they headed for a swimming spot. The prisoners sang "Wir Fahren Gegen Engeland," or, roughly translated, "We Are Off to Conquer England," which raised the ire of the Norwegians. The flyers had just learned how to execute an aerobatic roll. They bought a couple of sacks of potatoes and each pilot took about 50 potatoes in his aircraft. As they flew low over the marchers, the pilots turned their planes upside down, and the potatoes fell like rocks on the men below.[5]

The relationship between the Norwegians and the Veterans Guards, who provided security at Camp 20, was a positive one. Per Conradi Hansen wrote that, as the guards were Canadian Army veterans of World War I, "they were a great group of warriors and a pleasure to meet and many a mug of beer was downed while we reminisced about our experiences."[6]

To help feed the men in Camp 20, farm property was leased on Jones Road, northeast of Gravenhurst, about five kilometres from the internment camp. German prisoners, under escort, walked back and forth to the farm daily during the growing season, where they planted vegetables, and raised pigs and chickens.

The prisoners of war also carried out some work benefitting Gravenhurst, building a low stone wall at Gull Lake Park's beach, along with a beautiful set of steps leading from the street level to the beach.

Prisoners sometimes escaped from Camp 20 and the RNAF often assisted Canadian police and military forces in searching for them. On December 8, 1940, seven prisoners appeared to have broken out from Camp 20 by means of a snow tunnel, but six were recaptured within a couple of days. However, one prisoner remained missing: Lieutenant Siegfried Schmidt. RNAF members skied through the woods around Gravenhurst, hunting for him.

The German wasn't found for four months. As it turned out, he had never left the camp and had been hiding in an attic, planning to escape by himself when the searchers gave up. When he was located, authorities learned that he had altered his appearance to avoid recapture. A fellow prisoner of war, who was a doctor, had fashioned an improvised scalpel and with it, implanted a piece of hockey puck in Schmidt's chin. When searchers finally located Schmidt, his chin had swollen two inches beyond normal, according to the rather gruesome report in the *Gravenhurst Banner*, and "had the appearance of a golf ball stuck there, which drew the fleshy part of his face like stretched elastic, making his features altogether different." After capture, Schmidt received medical attention and survived.[7]

According to *The Gilded Cage*, all escapees from Camp 20 during wartime were recaptured, except one. That individual was Lieutenant Walter Manhardt, who made his way to the United States and lived there, concealing his past, for the rest of the war.[8]

Some of the prisoners of war suffered from tuberculosis and were taken to Muskoka Hospital, a Gravenhurst tuberculosis sanatorium, for x-rays and treatment. A number of RNAF trainees also received treatment for tuberculosis there. Muskoka Hospital staff tried to avoid having Norwegian personnel and German prisoners of war present for treatment at the same time, by scheduling their appointments to take place on different days. However, sometimes these plans fell apart, and when the Germans and Norwegians met at Muskoka Hospital, the atmosphere seethed with tension.[9]

Jobs at Little Norway

As well as Norwegian staff, Little Norway employed some local citizens in a variety of roles at the camp. Christine Cator Everest recalls her mother, Mary Cator, and her aunt, Vera Wallace, working as cleaners there. They travelled back and forth each day by bus from Gravenhurst's post office to the camp.[10]

Frank Laycock, later a well-known Gravenhurst milkman,

worked as a labourer at Little Norway. His tasks included emptying garbage pails from the camp's offices and burning the trash. One day Laycock was shocked to hear gunshots coming from very close at hand. Someone had dumped live ammunition into a garbage pail, the garbage pail had been emptied into the blazing fire, and the ammunition was exploding. Luckily, Laycock wasn't hurt, but Second Lieutenant Rolf Lund was very unhappy. He lined all RNAF members up for a stern lecture on safely handling ammunition.[11]

Norwegian Women's Auxiliary Air Force

In late 1943, the Norwegian Women's Auxiliary Air Force (NWAAF) began recruiting women, who were either Norwegian citizens or Canadian citizens born in Norway, to join their ranks. A similar organization already existed in England.

Generally, women wishing to serve with the NWAAF worked at Little Norway for a probationary period in jobs such as secretary, kitchen helper, cook, airplane painter, hospital helper, and waitress. After their probation finished, the young women enlisted in the NWAAF, underwent training, and then usually returned to their jobs.

The first course started on September 29, 1943, when 11 women spent three weeks practising drill and studying military regulations and procedure. Three of these courses were given over the war.

The NWAAFs had no officers of their own, but reported to male RNAF officers. The women didn't live at Little Norway, but commuted to the camp from their homes daily.[12] Some NWAAFs were married to members of the RNAF, including switchboard operator Elsa Erland.[13]

Airmen repairing aircraft's engine.

PHOTO: LITTLE NORWAY MEMORIAL.

A jacket, wedge cap, and four-pleated skirt made up the uniform; one woman told a *Toronto Daily Star* reporter that she liked the skirt as it allowed more freedom for marching and walking than did a straight skirt.[14]

By the end of the war, 27 women had enlisted in the NWAAFs in Canada and some 50 had enlisted in England.[15]

Gravenhurst nurses and local volunteers at Blood Donor Clinic, Little Norway. To name a few: bottom row, from left, Edna Groh, Eve Crawford, and Mad Hall. Second row, from left, Katherine Stratford, Helen Guest. Third row, at right in a polka-dotted dress, Peg Powell. Fourth row, at right, Betty Davis. Fifth row, at right, Ellen Drury. PHOTO: GRAVENHURST ARCHIVES.

Assisting in patriotic parades and drives

Muskoka's towns enthusiastically supported Canada's war effort. Towns organized scrap drives where they collected metal, rubber, and other items to be recycled and used to build aircraft, ships, and tanks. Citizens became accustomed to the rationing of products like sugar, tea, coffee, meat, butter, gasoline, and tires.

Muskoka held a series of Victory Loan Drives over the course of the war and always surpassed its targets. The "Victory Train," a 15-car train carrying equipment and representatives from the Canadian Armed Forces, stopped in Gravenhurst and Bracebridge in February 1942. It was travelling across Canada to support the Victory Loan initiatives and to stimulate recruiting. Many citizens and most local schoolchildren visited the train.[16]

Little Norway helped out with the local wartime drives. Individual airmen and airwomen subscribed for Victory Bonds and took part in blood donor clinics.

When Gravenhurst's Rotary Club organized a squadron of Air Cadets in 1943, the Norwegians provided a large military bus, free of charge, to assist with transportation. The bus took the young men to and from their meetings, which were held at Massey Hall on the grounds of Muskoka Hospital.

On August 6, 1943, Coca-Cola Company Limited of Canada put on a variety show in the hangar at Little Norway; this was one of the "Victory Parade" performances sponsored by the company at military camps and war plants across Canada. Jack Brown, Coca-Cola bottler in Gravenhurst, hosted the event.

The show featured Mart Kenney and his Western Gentlemen, a popular Canadian band. Cy Mack, a CBC radio announcer, was Master of Ceremonies; some 25 minutes of the show were transmitted live over the CBC broadcasting network. After the stage show, the guests toured Little Norway and enjoyed watching Funny, the bear, drink Coca-Cola after Coca-Cola.[17]

The RNAF also participated in local military events. For

Victoria Day in 1943, George Panter, a Gravenhurst World War I veteran, organized a salute to the Veterans Guard. Events began spectacularly, with three flights of Norwegian aircraft flying in formation and performing breathtaking manoeuvres above a crowd of over 2,000 people gathered along Muskoka Road in Gravenhurst. It was the first time an air show had taken place in the town and many came from neighbouring localities to enjoy the display.

This was followed by a giant parade, composed of the Gravenhurst Bantam Brass Band; three platoons of Veterans Guard; RNAF units carrying a huge Norwegian flag and banner; Huntsville Sea Cadets and Orillia Air Cadets; representatives of the Royal Canadian Legion branches in Gravenhurst, Orillia, Bracebridge and Huntsville; Boy Scouts and Girl Guides. After speeches by local dignitaries, each armed forces group demonstrated military drills to an appreciative audience in front of the post office.[18]

Operational changes at camp

On September 15, 1943, Captain Øistein Røed replaced Captain Hans Lund as Little Norway's commanding officer. Also that year, air crew taking classes at the BCATP Service Flying Training School studied in Ontario rather than in Moose Jaw, Saskatchewan, or Medicine Hat, Alberta. For single-engine aircraft, student pilots took lessons at Camp Borden, located near Barrie; for twin-engine planes, at Hagersville, southwest of Hamilton.

With the RNAF's move from Toronto, Vesle Skaugum became the new site of the Radio School.

Although the Navy Air Service and Army Air Service had been operating jointly for some time, the two were officially made one air force, the Royal Norwegian Air Force, by Royal Proclamation on November 10, 1944.

Accidents

More fatal accidents occurred both before and after Little Norway moved to Muskoka.

Private Nicolay S. Selmer died on January 9, 1943, when his twin-engine Avro Anson crashed at Moose Jaw, Saskatchewan.[19]

Blood Donor Clinic at Little Norway. Royal Norwegian Air Force airmen donate blood, assisted by nurses Mad Hall and Katherine Stratford.

PHOTO: GRAVENHURST ARCHIVES.

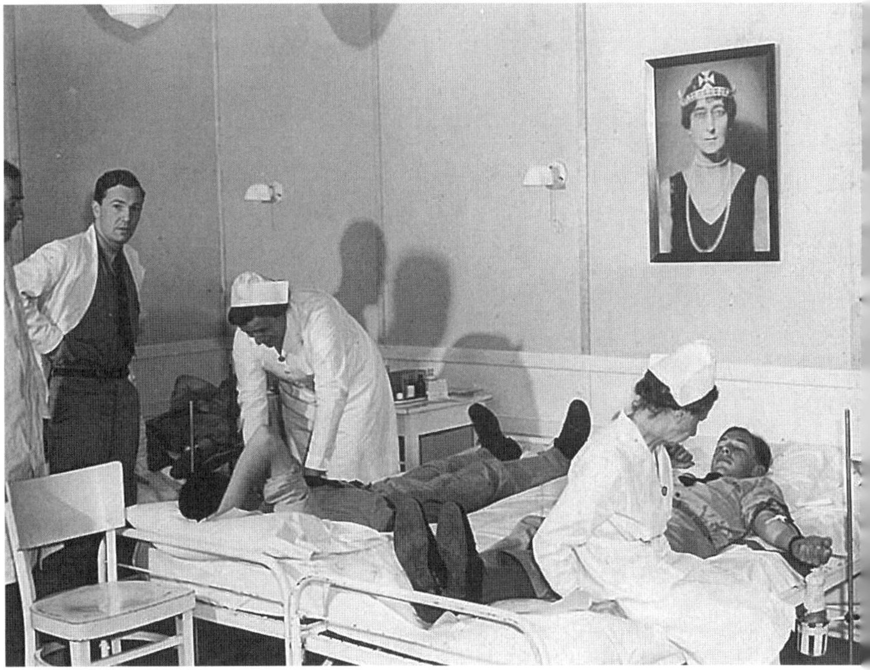

Royal Norwegian Air Force airmen at Gravenhurst Train Station.
John Stene is sixth from the left.

PHOTO: LITTLE NORWAY MEMORIAL.

Sergeant Ole Bacher, 21 years of age, was returning from a routine training flight with six other pilots on January 25, 1943, when the group encountered an unexpected blizzard over Toronto. Sergeant Bacher's aircraft became lost and crashed in flames at a farm near Brampton, killing him.[20]

On February 22, 1943, Second Lieutenant Conrad Mohr radioed to Toronto Island Airport, "I am now starting a power dive." His Curtiss aircraft never pulled out of the dive, plunging to earth and exploding with a tremendous sound, heard by thousands in downtown Toronto.[21]

Another fatal crash occurred in Muskoka on August 26, 1944. Sergeant Roy C. Heise and Airman B.O. Ottersen died instantly when their Fairchild training plane crashed in a swampy area near Germania, a small settlement about 10 kilometres from Muskoka Airport. During aerobatics, a wing broke off the aircraft and neither man was able to parachute to safety. Heise, a Swede, had battled the Soviets in Finland and then fought the Germans in northern Norway before escaping to Canada.[22]

A similar crash occurred on October 16, 1944, near Reay Lake. Two instructors, Lieutenants S. Sandvik and O. Sælid, were practising instrument flying in a Fairchild training plane when one wing broke off the aircraft. Both trainees parachuted out, Sælid landing safely but Sandvik suffering a split kneecap. Getting Sanvik out through the thick brush was very difficult, and it was impossible to extricate the whole aircraft, so the RNAF salvaged what they could and left the rest where it lay. In recent years, some of the wreckage was discovered in the bush near Reay Road.

A structural fault had contributed to both of the latter accidents. After making wing modifications, no further mishaps of the kind occurred.[23]

March of Time Fairchild, Little Norway.

PHOTO: LITTLE NORWAY MEMORIAL.

26

Heavy Water

At the site of the Norsk Hydro Electric plant in Vemork, Norway, a hydrogen factory produced "heavy water," a special type of water containing the hydrogen isotope deuterium. Heavy water was used in the making of nuclear weapons.

The Allies knew German scientists were working towards development of the atomic bomb and that Germany's occupation of Norway gave it access to the Vemork heavy-water operation. Possession of the atomic bomb would likely have meant victory in the war for Germany, and so in early 1942, the Allies made destruction of the Vemork plant their number one priority.

The SOE sent a Norwegian secret agent to the area in March 1942, and, in October, he was joined by four more agents, mostly from Norwegian Independent Company No. 1. This group's task was to prepare for the arrival of 34 British saboteurs, scheduled to fly into the area in two gliders; the saboteurs would then blow up the plant. However, the plane bringing one glider collided with a mountain, and both it and its glider were destroyed. The other glider also crashed. Those passengers who survived the accidents were killed by the Germans.

Six more secret agents arrived by parachute on February 16, 1943, to join the original unit. On February 27, nine of the agents climbed down a steep, rocky gorge after dark, crossed a river, and hiked up a steep hill on the other side to reach the Vemork plant. Using wire-cutters, the men made their way

Vemork hydroelectric station, Norway. During World War II, "heavy water" was produced in the rectangular, multi-storey factory building in front, and electricity in the classical building to the rear. Norwegian saboteurs blew up the heavy water factory on February 27, 1943. PHOTO: NORSK HYDRO PHOTO COLLECTION / NORWEGIAN INDUSTRIAL WORKERS MUSEUM, NO. 4F01326.

through the plant's outside gates. The agents set up explosives inside the plant and then skied back to safety.

The subsequent blast destroyed much of the plant and the heavy water stored there. The mission was a great success for the SOE and Norwegian Resistance.

However, after a few months the Vemork plant had been rebuilt and began producing heavy water again. The Americans bombed the heavy water plant and the power station in November 1943.

To ensure the continued production of heavy water, the Germans decided to move the equipment and some semi-processed heavy water to Germany. The Norwegian secret agents, who had remained in the mountains after the Vemork plant detonation, received orders to attack a ferry, which was carrying the heavy water and equipment across nearby Lake Tinnsjø. On February 20, 1944, the agents' explosives blew up the ferry and it sank; four Germans and 14 Norwegian civilians who were aboard died.[1]

Oluf Reed-Olsen, Gunnar Sønsteby, and Max Manus

After parachuting into Norway in spring 1943, Oluf Reed-Olsen and his colleagues had established secret wireless stations in countryside locations near Kristiansand and later in Lommedalen, near Oslo. They were sending intelligence about German aircraft and shipping to Britain, as well as carrying out other espionage missions. In early 1944, Reed returned to England to meet with officials. While there, he was given two months' leave and a free trip to wherever he wished to go. He returned to Toronto on the QUEEN MARY. *Ten days after he arrived, he married Joan Whiteside.*

In April 1944, Reed returned to the British Secret Service. At

Joan Whiteside marries Oluf Reed-Olsen,
St. Aidan's Church, Toronto, February 24, 1944.

PHOTO: OLUF REED-OLSEN FAMILY COLLECTION.

Tempsford, England, on May 6, 1944, he once again found two eggs on his supper plate. That night, he and another agent parachuted back into Norway and resumed their radio transmission work and espionage.[2]

Reichskommissar Terboven and the Nasjonal Samling continued to take strong measures to tighten control in Norway. In early 1944, the Nasjonal Samling's Minister of Justice wrote to the German SCHUTZSTAFFEL *(SS), the Nazi paramilitary organization, suggesting that 75,000 Norwegians, aged 18 to 25 years old, should be registered and made available to fight the Soviets on the Eastern Front with the SS.*

When the Resistance secretly got a copy of the letter, they carried out a number of attacks against the labour registration offices. In one instance, Gunner Sønsteby, Max Manus, and another agent, Gregers Gram, blew up the Oslo labour registration office the day before the young Norwegians were required to register for the labour draft. No one was killed. The Germans had to postpone the draft.[3]

The Resistance sent appeals, through the underground press and through the BBC, *asking people not to register for the labour draft. In the end, about 30 per cent of those required to register didn't report in the first round of registrations. In the second round, the great majority didn't register, and many of those young Norwegians went into hiding in the mountains and forests.*

A few months later, the Nasjonal Samling announced that new ration cards wouldn't be given to those who hadn't reported for the labour registration. This was a serious problem as many Norwegian families would have starved without the ration cards.

The Resistance decided the best approach would be to hijack a truck filled with new registration cards as it left the printers to drive to the post office. On August 9, 1944, Gunnar Sønsteby and two other agents carried out the mission, commandeering a truck containing some 150,000 new ration cards. After transferring the cargo to a truck of their own, they let the police know the Resistance had the ration cards and was, in effect, holding them for ransom.

The Nasjonal Samling surrendered, as the unauthorized use of the stolen cards would have created chaos in the administration of the country's rationing system. The government gave up on the idea of withholding ration cards from those who hadn't registered for the labour mobilization initiative. Then, having won the battle, the Resistance returned the ration cards to the government.

In late 1944, Sønsteby was made sabotage leader for the Central Committee of the Home Forces, performing missions as ordered by that committee or by London directly. His group of saboteurs was known as the "Oslo Gang."[4]

Police Force

In 1943, thousands of young Norwegian refugees remained living in Sweden, working in the forests and living in refugee camps. To help prepare for Norway's eventual liberation, the Norwegian authorities set up camps across Sweden, where the refugees practised drills, learned to use weapons, and gained other military skills. By the spring of 1945, some 13,000 stood ready as a "police force," intended to serve as peace-keepers when Norway gained its freedom.[5]

27

Norwegian Flyers Overseas

By late 1943, the Allies seemed to be gaining the upper hand in Europe. Air crew and ground crew from Little Norway contributed to victories over Nazi forces.

John Stene

John Stene's first mission with Bomber Command, flying a Halifax aircraft through Germany's Ruhr Valley, was one of his most dangerous experiences.

The Ruhr Valley, he said, was "one enormous hornet's nest of guns and search-lights." Flak exploded everywhere. When John's bomber became caught in a huge cone of searchlights for several minutes, anti-aircraft ground fire concentrated on the aircraft and hit several fuel tanks, leading to a hazardous loss of fuel.

However, the aircraft managed to return to the nearest airfield in south England. After landing, the air crew discovered that the German guns had left nearly 30 holes in the aircraft's body and the oil tanks for two of the four engines were nearly empty. Those two engines would have stopped midair if they'd been running any longer.

John would later recall that he wasn't really scared or seriously concerned when the attack was at its worst, but added that this might have been because he didn't know any better.

Subsequently, John and his air crew flew bombing missions to Berlin, Stuttgart, and other cities in Germany; they also bombed locations within France. While the enemy flak and searchlights inspired awe in the RAF flyers, John believed the Luftwaffe's night fighters, with their efficient aircraft and highly skilled crews, posed the greatest danger.

Only 25 per cent of the air crew in Bomber Command could hope to survive an operational tour of duty, which amounted to about 30 trips. It was vitally important for the air crew to work as a team, with each man knowing he could rely completely on the other. John believed that if a crew could complete their first six to eight trips safely, the experience gained working together would greatly improve their chances of survival on the remaining missions. Also, the air crew's skills needed to remain honed, which was easiest when the men flew about four to six trips every month.[1]

Still, chances were slim. One former pilot said that you couldn't afford to think about losses too much:

> *It was the same even if we lost our best friend. In my group only two of us who went overseas survived. All the rest were shot down.... It wasn't that you weren't hurt. It was a terrible loss of course. But we had trained ourselves to phase it out and not think about it. If we had thought about that all the time we wouldn't have been able to fly!*[2]

John wrote that, somehow, the air crew managed to adjust to the situation, in spite of the great strain of seeing friends disappear, one after the other. He said, "We knew we were all volunteers and that

Halifax bomber.

PHOTO: UNITED KINGDOM GOVERNMENT.

BELOW Martin Gran, Norwegian No. 332 Squadron. Gran painted a swastika on his Spitfire every time he shot down an enemy aircraft. Over the course of the war, he would be credited with destroying nine. SOURCE: KÅRE FASTING: *FRA KONTRASKJÆRET TIL TOKYO*, OSLO, 1959.

we had taken on a job which had got to be finished and that in this game we were all 'playing for keeps.'"

In John's mind, they were living a double life. In one world, the air crew had comfortable quarters, good friends, and leave every sixth week; in the other world, they took on dramatic, intense, life-threatening danger where the outcome was always unknown and luck was the most important ingredient.[3]

With such great losses, officers got rapid promotions, and within a period of three to four months, John went from lieutenant to captain to major.

Squadron 331 and 332

As the Allies experienced gains against Germany in Italy and the Soviet Union, they began planning for an invasion of continental Europe. There they would face fierce opposition, as Germany had heavily fortified the European coastline with artillery, mines, and other obstacles to an invasion by sea. The stretch between Denmark and southern France was known as "Fortress Europe."[4]

Preparations for the invasion, called "Operation Overlord," began long in advance of the event. On November 1, 1943, Squadrons 331 and 332 were combined into 132 (Norwegian) Wing of No. 84 Group in the Second Tactical Air Force. British RAF Squadrons 66, 127, and some other groups joined 132 Wing and in spring 1944, the entire group moved to Bognor, on England's south coast.[5]

Lieutenant Helge Mehre became wing commander in November 1943, and Major Wilhelm Mohr the wing's administrative leader. Captain Leif Lundsten led Squadron 331; Major Werner Christie commanded Squadron 332. About 70 per cent of the wing's personnel were Norwegian and 30 per cent British.

Operation Overlord was set for June 6, 1944, "D-Day," when soldiers from Britain, Canada, the United States, and other countries were to land from troop ships at five beaches in Normandy, France. On the evening of June 5, 1944, Lieutenant Mehre gathered the wing's pilots together at their Bognor base and said, "Well boys, this is it!" then informed them of the next day's planned Allied invasion of occupied France.[6]

The wing's task would be to gain air superiority and protect the Normandy beaches from enemy aircraft while Allied soldiers went ashore; over the course of D-Day, the wing would fly four sorties.

Ground crew had painted white and black stripes on the wings and fuselages of the Spitfires to mark them as Allied aircraft and help avoid attack by friendly artillery.

In the darkness, the Norwegians watched in awe as the invading aircraft passed overhead on their way to France. Some 4,000 Allied bombers, along with 3,700 fighters and fighter bombers, would take part in Operation Overlord.[7] Wilhelm Mohr wrote: "The thunder from transport aircraft with paratroopers and gliders dragging behind could be heard throughout the night of 5 June. It was like a carpet of aircraft, but still the silhouettes of single aircraft could be seen against the clear night sky."[8]

Wing 132 expected to encounter massive Luftwaffe resistance on the first day of Operation Overlord, but actually saw no enemy aircraft at all. The German military leadership hadn't thought an invasion would take place in Normandy, or that any major invasion was imminent, and so was unprepared for the vast Allied landings of June 6, 1944.

Troops of the 9th Canadian Infantry Brigade (Stormont, Dundas, and Glengarry Highlanders) going ashore from LCI (L) 299 (Landing Craft Infantry), Bernières-sur-Mer, Normandy, France, 6 June 1944. PHOTO: GILBERT ALEXANDER MILNE / CANADA DEPT. OF NATIONAL DEFENCE / LIBRARY AND ARCHIVES CANADA / PA-122765.

Below the aircraft, 7,000 ships of all kinds took part in the assault. On the five fortified beaches, troops began to come ashore, including 14,000 Canadian soldiers at Juno Beach. They faced brutal artillery fire, mines, razor wire, and other dangers as they landed and fought their way forward. Around 3,000 Allied troops died on D-Day, and 340 of them were Canadian.[9]

Little Norway's airmen listen to radio broadcasts about D-day landings, June 6, 1944. Svend Clement is second from left.

PHOTO: LITTLE NORWAY MEMORIAL.

Rolf Kolling remembers flying two sorties on D-Day and looking down from his Spitfire at the multitude of ships unloading troops, who were wading ashore. He watched as German ground fire destroyed a big British troop transport before any of the soldiers could disembark from the boat. Kolling later told a reporter: "It was awful. I felt like a spectator."[10]

The Luftwaffe soon joined the battle in the days following D-Day, as Allied troops continued to land and fight through terrible conditions. Wing 132 flew patrols over Normandy and, on June 7, lost one of their own to German flak. Two days later, friendly fire took the life of Squadron 331's Commanding Officer, Major Leif Lundsen, when his aircraft was accidently shot down by the Royal Navy.[11] Other losses followed.

On June 15, Rolf Kolling's plane was nearly destroyed by German gunfire. A shot had hit the aircraft, travelled under Kolling's left leg, and struck the fuse box. Kolling was unharmed, but he had the very difficult challenge of flying the Spitfire and landing it without the use of instruments.

He managed to make an emergency landing at a small airport in northern France. Airport staff examined the plane and told Kolling that the gunfire had so damaged the Spitfire that it would never fly again. He felt very lucky to have survived the attack, but then wondered how he would return to England. Fortunately, a Douglas DC-3 air ambulance happened to be at the airport at the time and it gave Kolling a lift to London. Within a couple of days, he was back flying with Squadron 332.[12]

As the Allies fought their way forward into France, Wing 132 followed, flying from airfields on French soil and attacking German defences. After Allied troops had liberated a landing strip, RNAF ground crews and their equipment arrived next, ready to support the Norwegian aircraft that would be flying from the landing strip. Sergeant Per Ole Falkendal Nilsen, flight mechanic, was one of the ground crew who landed at Normandy around August 22, 1944, and travelled to Caen airfield to prepare for the arrival of Wing 132's Spitfires.[13]

The Allies liberated Paris on August 25, 1944, effectively bringing Operation Overlord to an end. Wing 132 moved

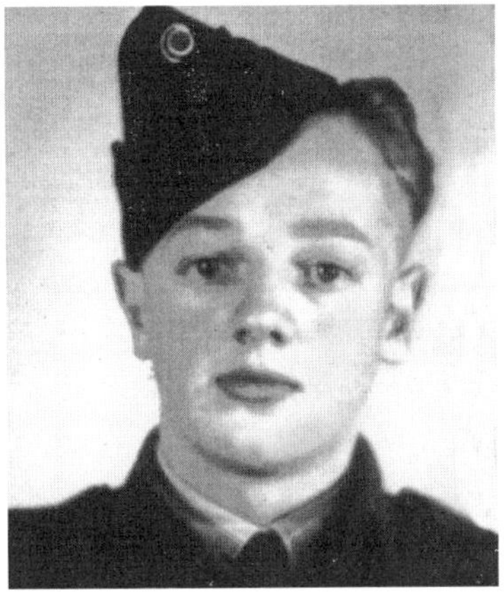

LEFT

Per Ole Falkendal Nilsen.

PHOTO: EVA FALKENHALL COLLECTION.

RIGHT

Norwegian airmen on the continent. PHOTO: NATIONAL ARCHIVES OF NORWAY.

northeast with the Allies, providing air support as the troops battled forward, participating in the liberation of Belgium and the Netherlands and attacking locations in Germany.

Wing 132's last base in continental Europe was at Twente, Netherlands; its last day of operations was April 21, 1945. Squadrons 331 and 332 were transferred back to North Weald, England, and subsequently to Dyce in Scotland.

Squadron 333

During the summer of 1943, Per Conradi Hansen was posted to Squadron 333 B Flight, at Leuchars, serving as a navigator on Mosquito bombers. While flying with pilot Lieutenant Jacob Martinius Jacobsen over the North Sea on June 16, 1944, they spotted a U-boat. During the succeeding battle, the U-boat was severely damaged, but it managed to shoot down the Mosquito. Both Norwegian airmen survived but with injuries, and spent the next 32 hours in a leaking dinghy on the cold ocean before being rescued by German forces. Hansen and Jacobsen then became prisoners of war in Germany for the rest of the war.[14]

Squadron 333 A Flight hunted for German U-boats, escorted convoys, landed secret agents in Norway, picked them up again, and flew reconnaissance missions. Rolf Hauge flew on A Flight's Consolidated Catalinas; in fact, the Catalina was Hauge's favourite plane, even though the unheated aircraft could get extremely cold when flying in Arctic areas.

Hauge was serving as an air gunner/mechanic on a Catalina flown by Lieutenant Carl Krafft on June 17, 1944, when the air crew spotted a U-boat on the surface of the ocean, northwest of Kristiansund. Taking advantage of cloud cover, the Catalina swooped down on the unsuspecting U-boat. The Catalina destroyed the German vessel with depth charges, and Squadron 333 was credited with sinking the submarine.[15]

A week later, Flight Lieutenant Dave Hornell from Mimico, a suburb of Toronto, was piloting a Catalina and returning to the RAF's Iceland base after a long patrol over the Arctic Ocean. The crew spotted a German U-boat on the water's

Members of Norwegian No. 333 Squadron, B Flight, Leuchars, Scotland, 1944. Jakob Anonby wrote the names of the men on the photograph as follows: from the left, L. Eirvåg, Andreassen, Lofnes, Turnbull, Johansen, Gabrielsen, Toby, Sven, Mollan, Anonby. PHOTO: ANONBY FAMILY COLLECTION.

surface, which fired on the aircraft and managed to strike its engine. Before crashing, the Catalina launched four depth charges at the U-boat and it sank, quickly. All eight members of the Catalina crew lived.

The Catalina had two four-person inflatable life rafts for the survivors to use, but one exploded, leaving one life raft to be shared by eight. The airmen took turns, one or two of them floating in the icy water for a time, hanging onto the side of the life raft, until others took their places. The wind was cold and waves washed over the life raft, so the men worked all night to stay awake, singing, shouting, slapping each other in order to keep from freezing to death.

In the dawn, an aircraft appeared in the distance. The men in the life raft desperately tried to attract its attention. They had three flares. They fired the first two to no avail, but the third flare caught a crewman's eye and the plane turned towards them. This was Catalina D-FP 183, called *Vingtor II*, on a routine patrol from the Shetland Islands—and its crew included Rolf Hauge, who spotted the flare.

The water was too rough for *Vingtor II* to land, and the high winds made flying very difficult; however, the aircraft stayed with the men in the life raft for 15 hours, circling above it.

After a time, a second aircraft arrived to help. The air crew tried to drop a life boat to the men in the water but, unfortunately, they dropped it on the wrong side of the life raft and it drifted away.

Vingtor II kept up radio contact with other aircraft in the area, and was able to guide a rescue ship to the life raft's location. Finally, nearly out of fuel, *Vingtor II* returned to its base, and a Sunderland aircraft replaced it.

When the RAF rescue ship finally arrived, the men had been in the life raft for 21 hours, and two had died. Pilot Hornell, now blind and unable to move, died on the rescue ship. The other crew members survived.

It's believed that Hornell died due to greater exposure to the elements, having given up his place in the life raft to another man. Hornell was awarded a posthumous Victoria Cross.[16]

Squadron 334

On May 26, 1945, Squadron's 333's B Flight moved to Banff, Scotland, and was designated as Squadron 334. Its Mosquito bombers hunted for enemy ships and U-boats along the Norwegian coast.

Stockholm route

In 1944, Bernt Balchen set up an operation flying American Lockheed Lodestars back and forth between Stockholm, Sweden, and Leuchars, Scotland. Aircraft on this "Stockholm Route" flew over occupied Norway in the dark, bringing Norwegian refugees and cargo to the United Kingdom and, in return, carrying ammunition, medical equipment, and other strategic war supplies to Sweden.

Finn Thorsager was among the pilots assigned to fly Lodestars on the dangerous Stockholm Route until the war ended.[17]

Hitler's forces retreat in Northern Norway

Meanwhile, Finland had been battling the Soviet Union since 1941 in a conflict called the "Continuation War," and had received weapons and military support from Germany. Finland and the Soviet Union reached an armistice in September 1944, and, as part of the peace, Finland agreed that the German troops had to leave the country.

The Soviets drove the Germans out of Finland, and Hitler's forces retreated through Finnmark, in northern Norway. To delay the Soviet advance, the Germans adopted scorched-earth tactics, burning homes, stores, and boats, and demolishing infrastructure and anything else that might be useful to their enemy. The Germans also forced most of the Norwegian population to evacuate the area and move to the south.

The displaced Norwegians endured great hardships, losing their homes, livelihood, and sometimes their lives. But the very fact that Germany was retreating from northern Norway brought hope. The Government-in-exile called it "the commencement of Norway's liberation." In a press release, the Government-in-exile sent warmest greetings to the Soviets, assuring them that the Norwegian Armed Forces and people would "do their utmost in the common struggle against the German oppressors."[18]

Northern Norway in ruins during the German retreat. This photograph was taken in the centre of Kirkenes, where destruction was almost total.
PHOTOGRAPHER: OLE FRIELE BACKER. NATIONAL ARCHIVES OF NORWAY.

In late October 1944, members of the Norwegian Army, who had been training in Britain, joined the Soviets in battling the Germans in Finnmark. Local volunteers entered the conflict, and in January 1945, their numbers were reinforced by the Norwegian "police force," refugees who had been secretly training in Sweden's camps for the previous two years.

FAREWELL TO CANADA

28

Farewell to Canada

By the end of 1944, with the Allies' victories following D-Day, the Soviet successes on the Eastern Front, and the German retreat in northern Norway, it seemed as if World War II would soon be over.

The BCATP began to wind down its courses in late 1944, and closed operations on March 31, 1945.

The RNAF leadership took steps to prepare to return to Norway following liberation. It reduced the numbers of air crew training in Canada and decided to transfer all personnel to Winkleigh, England. As most of the recent recruits had come to England from Sweden, it was more cost-effective to train them right in England than to send them to Canada.[1]

But before the RNAF left Canada, a special ceremony took place at Little Norway to say farewell.

Knots of friendship

On February 17, 1945, Little Norway officially closed. Crown Prince Olav and Crown Princess Märtha travelled to Gravenhurst with their attendants for the events, and stayed for two nights at Martin Manor, a lodge on the main street. When first arriving in town, the Royal party was hungry and hoped to find a restaurant open for a quick lunch. However, all restaurants were closed and the town was nearly empty. Most citizens had already left to attend the afternoon ceremony at Little Norway.[2]

Over 1,000 guests from Toronto, Ottawa, Gravenhurst, Bracebridge, Huntsville, and other spots gathered for the farewell, including dignitaries like Lieutenant Governor Albert Matthews; RCAF Air Vice-Marshal W.A. Curtis; Muskoka Member of Parliament Stephen J. Furniss; Norway's Consul General, Daniel Steen; and mayors Robert Saunders from Toronto, John Gibson from Gravenhurst, E.A. Whitten from Bracebridge, and H.E. Rice from Huntsville.

A special train carried some 500 people from Toronto to Gravenhurst and buses delivered the crowd from the train station to Little Norway. One of those people was young Rita (Abel) Hiorth, who had darned socks for the airmen with her mother and the other ladies when the RNAF was stationed in Toronto. Rita remembered the bitterly cold day, and a big

OPPOSITE Crown Prince Olav speaks to the guests at the February 17, 1945 "Farewell to Canada" ceremony at Little Norway. To the left of the stage, sits the RCAF's Central Band; to the right, the combined choir from Little Norway and Gravenhurst's United Church.

PHOTO: KAREN GREAVETTE COLLECTION.

Little Norway officially bids farewell to Canada at a ceremony held in the camp's aircraft hangar on February 17, 1945. Lieutenant Colonel Ole Reistad addresses the guests. Crown Princess Märtha and Crown Prince Olav sit behind the table, in the first row. PHOTO: LITTLE NORWAY MEMORIAL.

crowd of people gathered in Little Norway's hangar awaiting the closing ceremony.[3]

Nine Fairchild aircraft flew in formation above the camp, thrilling the guests. Things had nearly gone very wrong. Spectators gasped as two planes clipped wings and pieces of broken wing fell to the ground. Luckily, both aircraft landed safely, with only minor damage. However, this close call served as a reminder of the dangers that air crew always faced.

The stage had been colourfully decorated with Canadian, Norwegian, and American flags; above them were posted the words "Farewell to Canada" and a depiction of wings. While the assembled guests felt great relief that the war was coming to an end, the day was tinged with sadness at the Norwegians' departure.

Events opened with music by the RCAF's Central Band from Ottawa. Lieutenant Oluf Reed-Olsen, who had escaped from Norway to safety in the United Kingdom during December 1944, was Master of Ceremonies for the two-hour event.

The first speaker was Leading Aircraftman E. Cramer-Gundersen, representing all the Norwegian airmen, who thanked Canadians for their kindness during the war years.

Lieutenant Colonel Ole Reistad told the guests:

> For five years we have been longing for the day the order should be given to break up camp and move eastward.... But Canada will forever occupy a special place in our hearts for the care and unique hospitality that has been shown us here.
>
> As we travel eastward now to be as close to home as possible when the hour of battle strikes, we send you all our heartiest and most deep-felt thanks.... Knots of friendship that we know shall be stronger till the day those at home will hear about the care and support that was given their flying sons on foreign soil in a dark hour.[4]

Crown Prince Olav expressed his gratitude to Canada and the RCAF for the great assistance given to the Norwegian airmen and women over the war years. Other speakers included Lieutenant Governor Albert Matthews, who regretted that Ontario had to part with the "gallant fellows" of the RNAF.[5] George Panter of Gravenhurst read a greeting from Canada's Prime Minister Hon. W.L. Mackenzie King.

Gifts were exchanged. Gravenhurst, Bracebridge, and Huntsville combined to present the RNAF with a large oil painting by famous Canadian artist A.Y. Jackson, entitled *Snow in October*. Toronto's Mayor Saunders gave the Crown Prince and Crown Princess an engraved silver tray. The Gravenhurst Rotary Club President, Lloyd Cosby, and the Gravenhurst Legion President, Hugh Pierre, jointly gave Crown Princess Märtha a solid gold maple leaf brooch, studded with a diamond. Mrs. Albert Matthews presented Crown Princess Märtha with a silver maple leaf set in pearls on behalf of the Imperial Order of Daughters of the Empire.

In return, the RNAF bestowed the first flag it flew at Little Norway in Toronto on Mayor Saunders, and gave a Norwegian flag to each of the mayors of Gravenhurst, Bracebridge, and Huntsville.

A combined choir from Little Norway and Gravenhurst's United Church, directed by Gordon Sloan, provided choral selections. While the *Gravenhurst Banner* called the singing "perfect," Cecil Porter, then a young teenaged member of the United Church Choir, thought otherwise. He, and others, had found it a real struggle to learn the words to the Norwegian anthem and he concluded that the audience probably found their rendition "memorable," but not in the way he'd hoped.[6]

At sunset, the guests gathered on the parade ground to watch the lowering of the Norwegian flag and the RCAF ensign, while the RCAF's band played. A plentiful smorgasbord followed in the messes, and guests had an opportunity to speak with the Crown Prince, Crown Princess, and other dignitaries. The guests also received a special, final issue of the camp's magazine, *Lille Norge Avisen*.

The evening ended with a dance in the hangar to the music of the RCAF's dance band. It was a colourful spectacle, with red-coated Royal Canadian Mounted Police, blue-uniformed members of the Canadian and Norwegian Air Forces, and other men dancing with "scores of smartly-dressed young ladies."[7]

Those heading back to Toronto caught buses from the camp to the Gravenhurst train station, where they took a special midnight train to the city.

Fire

Fire broke out in the officers' mess at Little Norway on February 21, 1945, around 5:30 a.m. Volunteer fire brigades from both Gravenhurst and Bracebridge responded to a call for assistance from the camp. The *Bracebridge Gazette* reported that Gravenhurst's Fire Department arrived first, but as it had

A fire broke out in the officers' mess at Little Norway on February 21, 1945. Half of the log building was destroyed.

PHOTO: KAREN GREAVETTE COLLECTION.

Airmen fighting the fire at Little Norway,
February 21, 1945, with water from the swimming pool.

PHOTO: KAREN GREAVETTE COLLECTION.

brought no pumper, the RNAF's pumper had to fight the blaze alone, by drawing its water supply from the swimming pool. In addition, airmen set up bucket brigades to help fight the fire.

The Bracebridge Fire Department soon arrived with its more powerful pumper. However, RNAF men had to shovel a six-metre-long laneway through the snow before the pumper could make a connection with the swimming pool. After this delay, the pumper poured water on the burning building and the fire was soon extinguished.

Half the big, beautiful, log building had been destroyed; nothing remained except brick fireplaces and chimneys. The officers' and NCOs' messes were lost. The kitchen and the other half of the structure, containing the airmen's mess, canteen, library, and some sleeping quarters, survived but suffered water damage.

About 200 RNAF personnel dined in restaurants in Gravenhurst and Bracebridge that day.[8]

Moving to England

On April 4, 1945, Lieutenant Commander Ole Reistad was transferred to RNAF headquarters in London, England, and Captain Ludwig Brannvoll took over as Little Norway's commanding officer.

Even though it would be some months before the RNAF's operation wrapped up in Canada, personnel and aircraft were gradually transferring to the United Kingdom.

As aircraft became obsolete or unneeded at Little Norway, they were sold back to the manufacturer or purchased by other governments. As an example, the Norwegians had received 36 Douglas attack bombers by 1941; after negotiations with United States and Canadian authorities, the Peruvian government bought 18. The U.S. War Department bought another 13 in 1943. Five were written off due to crashes.

When Little Norway closed, only Fairchild and Interstate Cadet airplanes remained. Seventy-five Fairchilds were flown from Little Norway to Toronto Island Airport, where they, along with two Interstate Cadet aircraft, were taken apart by ground crew and put in crates. The RNAF sent the crates in shipments to Winkleigh between January and April 1945, and directly to Norway after liberation.[9]

29

Victory

Hitler committed suicide on April 30, 1945.

Admiral Karl Dönitz, Hitler's successor, declared Germany's unconditional surrender on all fronts on May 7, 1945. Dönitz dismissed Josef Terboven as Norway's Reichskommissar and assigned Terboven's powers to General Franz Böhme, Germany's commander in chief in Norway.

May 8 was Norway's Liberation Day and the public celebrated joyously. That same day, Josef Terboven blew himself up with dynamite in a cellar under Skaugum.

Some 350,000 armed German troops remained in Norway and there was some uncertainty as to how they'd react. Would the Germans destroy buildings and blow up infrastructure as they'd done in the north? Or would Nazi forces come from Germany to Norway, to fight a last battle there?

General Böhme received orders from Berlin to capitulate. A group of Allied officers, led by British Major General Richard Hilton, flew to Oslo aboard one of Squadron 333's Catalinas, to meet with Böhme and receive the surrender of Germany's forces in Norway. Also aboard the aircraft was an air gunner, Sergeant Nils Christensen, who had trained at Little Norway.

The aircraft landed at Fornebu, and German soldiers conveyed its passengers to Oslo. Christensen recalled hearing sniper fire in Oslo, but people began to fill the streets as news of a possible German surrender circulated. He and the others were the first free Norwegians to return to Norwegian soil, and they opened the bus windows and handed out English cigarettes and candy. When the crowd realized who the men were, they became very excited and stormed the bus. Christensen and some others climbed onto the bus's roof, singing and cheering along with the crowd.[1]

Norwegians crowd Skatval train station on Liberation Day, May 8, 1945.

PHOTOGRAPHER: OLAV M. FISKVIK.

After General Hilton met with General Böhme, the latter capitulated effective midnight, May 8, 1945. As part of the surrender, Böhme ordered German soldiers to withdraw from Norwegian locations and to assemble in specified locations

Surrender of Akershus Fortress, Oslo, May 11, 1945. The German garrison's commander, Major Josef Nichterlein, and his aide, Captain Hamel, hand the fortress over to the Norwegian Resistance movement's Terje Rollem. PHOTO: NATIONAL LIBRARY OF NORWAY. PHOTOGRAPHER: JOHANNES STAGE.

where they would hand in their arms. The German troops obeyed Böhme's orders.

The Government-in-exile wished to ensure that law and order prevailed during this period of change, and so asked Milorg to take control of the country. Milorg, then 40,000 members strong, agreed. Its representatives occupied official buildings, including the Royal Palace. In its communications with the public, Milorg warned Norwegians against provoking their former enemy, especially while many Germans were still armed.[2]

Gunhild Ramm Reistad, who had been evicted with her family from their home in Oslo after the invasion so that a German official could live there, remembered how her 18-year-old brother walked up to the family home and knocked on its door on Liberation Day. He told the German who answered, that the family wanted their house back. The German agreed and moved out within a couple of weeks. Luckily, the house had not been damaged over the war years.[3] Gunhild Ramm would later marry Kjell Arne Reistad, one of Ole Reistad's children.

A new civil administration took over in Norway on May 9, arresting those who were believed to have collaborated with the enemy. Some 13,000 "police force" members, Norwegian refugees who'd been training in Sweden, arrived to support Milorg in keeping the peace. Allied soldiers assisted with disarming and returning German troops to their homeland.[4]

VE Day in Muskoka

When the news of Germany's surrender spread by radio on May 7, 1945, Gravenhurst erupted joyously.

Dorothy Shaw, Principal of Gravenhurst High School, told the students about Germany's surrender and that May 8 would be "Victory in Europe Day," soon to be better known as "VE Day." As Cecil Porter recalled, at almost the same time as Miss Shaw was speaking, a whistle began to blow at Rubberset Co. Ltd., a large paint brush factory, and the bell at the town fire department started to ring. "The school vacated within minutes," Porter wrote.[5]

About 100 Greavette Boats employees followed Bert Duncan, playing trombone, and Hugh Clairmont, on trumpet, in a spontaneous parade marching up Bay Street. Before long, at least 200 people from Rubberset had joined the parade, which had grown to a half-kilometre in length. High school students carried a large, improvised effigy of Hitler: a stuffed figure covered in white cotton, on which someone had drawn a Hitler-like moustache and swastika tattoos. The effigy was suspended between two long poles. It proved to be a popular attraction.

Cars drove very slowly along the main street, with happy young women perched on the cars' fenders or standing on their running boards.

For most of that sunny morning, citizens lined the main street, whistling, ringing bells, and waving flags. Stores closed, while town and church bells chimed and sirens filled the air. Many local people headed for Internment Camp 20, the German prisoner-of-war camp, where they stood near the barbed-wire fence and sang patriotic songs.

During the afternoon, the *Gravenhurst Banner* broadcast a program of stirring music and announced upcoming VE Day events from the newspaper's storefront window.

A Mosquito bomber roared over the town, flying low at a speed of about 640 kilometres per hour, momentarily shocking and deafening citizens. Its visit had been planned much earlier to promote the Victory Loan Campaign, but the aircraft added a thrilling touch to the day's celebrations.

That evening, Hugh Clairmont's Orchestra played at a public dance held by Mr. and Mrs. Joseph Chiarenza; Stanley's Orchestra supplied music for a street dance in front of the Gilmour Hotel. More solemnly, Trinity United Church held a special Thanksgiving service, which drew a big crowd.[6]

Huntsville, Bracebridge, and most other Muskoka communities enjoyed similar jubilant celebrations on May 7. During the afternoon, the Bracebridge town hall bell rang continuously for about a half-hour, "sometimes at the usual speed, sometimes at fire-alarm tempo," according to the *Bracebridge Gazette*.[7] There was a huge parade downtown, and spectators and marchers waved flags from nearly all the Allied countries, with British, American and Norwegian being the most numerous. The parade ended at Memorial Park, where Mayor Whitten spoke and a religious service took place.

Gravenhurst citizens celebrate the Allies' victory over Nazi Germany with an impromptu street parade, May 7, 1945.

PHOTO: GRAVENHURST ARCHIVES.

At night, Bracebridge citizens joyously paraded along central streets by torchlight while bands played. A truck joined in, carrying a homemade gallows, from which an effigy of Hitler hung; the effigy was burned at the end of Manitoba Street. The Citizens' Band played a concert and a street dance followed.[8]

May 8, VE Day, was a more formal day, set aside as a Day of Thanksgiving for the end of the European war. Events began in Gravenhurst with a Drumhead Service in Gull Lake Park, attended by some 2,000 people. Canon E.F. Pinnington led

Airmen from Little Norway take part in Drumhead Service in Gull Lake Park, Gravenhurst, on VE Day, May 8, 1945.

PHOTO: GRAVENHURST ARCHIVES.

the service, assisted by the RNAF's Lieutenant V. Neumann. Units from Little Norway, the Royal Canadian Legion, Veterans Guard, Air Cadets, Boy Scouts, Girl Guides, and Wolf Cubs all took part.[9]

A parade followed the Drumhead Service, enjoyed by enthusiastic, flag-waving crowds lining the streets. Afterwards, the crowd dispersed and in the evening, many people listened to special radio broadcasts relating to Germany's surrender.[10]

VJ Day

When Japan surrendered on August 14, 1945, Muskoka erupted in celebration again. In Gravenhurst, car horns honked, church bells rang, the Rubberset whistle blew, fire sirens sounded and people banged on tins and dish pans. Eric Ronald and Fred Girard, both playing drums, led a parade along Muskoka Road, with two American visitors performing majorette routines. The town band arrived, and car after car joined in as the procession grew to half the length of the main street. Even pouring rain couldn't stop the festivities, and Gravenhurst's citizens celebrated for two hours.

In Huntsville, the Anglo-Canadian Leather Company blew its whistle, car horns honked, and hundreds made their way to the town's main street at the news of Japan's surrender. Within a short time, the downtown became the scene of "wild but harmless jubilation."[11] In Bracebridge, fog horns, cheers, and drums added to the din as crowds gathered on Manitoba Street, which was decorated with streamers.[12]

The next day was Victory in Japan Day or "VJ Day," and the communities held Thanksgiving services, followed by parades.[13]

World War II was over.

30

Aftermath

The Norwegian airmen were finally going home. On May 22, 1945, 36 Spitfires flew from Dyce, Scotland, over the North Sea to Stavanger, Norway. Piloting one of these aircraft was Rolf Kolling. He remembers the wonderful feeling of seeing the Norwegian mountains rise again after spending the five years away, fighting the war.

The aircraft landed at Stavanger Air Base and Kolling telephoned his family, to whom he hadn't spoken during wartime. Kolling's mother answered the phone, and when he told her it was "Rolf calling," she asked, "Which Rolf?" as there were three men named Rolf in the family. When she realized the caller was her son, she nearly fainted.[1]

Per Ole Falkendal Nilsen returned to Norway in late May or June 1945, and discovered that he had two new siblings. As it had been so hard for RNAF members to communicate with family members in Norway during the war, Nilsen had no idea that his family had expanded.[2]

The termination of the war brought an end to military service for most RNAF members. Personnel stationed in England were discharged there; the handful remaining in Canada could choose to return to Norway on Norwegian ships, or to be discharged in Canada.[3]

King Haakon left England and sailed back to Oslo on the British cruiser *Devonshire*, the same ship in which he'd escaped from Norway. He arrived on June 7, 1945, exactly five years from the day he left, to rapturous crowds celebrating his return. Resistance fighters Gunnar Sønsteby, Max Manus, and others from the Oslo Gang served as bodyguards for Crown Prince Olav and King Haakon when they came home from England.

King Haakon's homecoming, Oslo, June 7, 1945.

PHOTO: OSLO MUSEUM, NORWAY.

The Oslo Gang received orders to pick up Vidkun Quisling from a police station and take him to Akershus for his trial

Vidkun Quisling in custody at Akershus Fortress.

PHOTO: NATIONAL ARCHIVES OF NORWAY.

on charges of high treason, murder, theft, and other offences. Quisling's personal circumstances had drastically changed. Gunnar Sønsteby described their meeting:

> Then there was our prisoner—the man we had hated and feared—powerless and helpless. We felt a little sorry for him. He fumbled with the blankets he had to take with him and dropped one of them. I felt that I wanted to help him pick it up.[4]

Vidkun Quisling was convicted of all the major charges brought against him and was sentenced to death. The Supreme Court rejected an appeal, and on October 24, 1945, Quisling was taken before a firing squad and executed at Akershus Fortress.

Ole Reistad

After Norway's liberation, Lieutenant Colonel Ole Reistad was assigned to Bardufoss Air Station as Commander of Air Command North, organizing the Norwegian air force there.

Tragically, Lieutenant Colonel Reistad developed a brain tumour and died at home in Oslo on December 22, 1949, at age 51. Before he died, Reistad was appointed a Commander with the Star of the Order of St. Olav, an award given for distinguished achievements on behalf of Norway and humanity.

Ole Reistad was highly respected by the people of Norway, who honoured his memory by naming many streets after him. In Elverum, the Ole Reistad Centre is the national base for glider flying. In Bardufoss, a cross-country ski race, called the "Reistadløpet," takes place annually.

War brides

Two hundred and thirteen Norwegians married during their time in Canada, and 93 children were born to RNAF personnel.[5]

With the end of the war, travel to Norway became safer for the young Canadian women who were married or engaged to members of the RNAF. Some 80, along with their children, sailed on the Swedish-American liner *Drottningholm*, which left on August 25, 1945, for Kristiansand.[6] About a month later, the liner *Stavangerfjord* carried 100 Norwegians, some with their Canadian wives and babies, from Halifax to Oslo.[7] The *Stavangerfjord* sailed again in late October, bringing another 60 war brides and their children to Norway.[8]

The women arrived in a country suffering the effects of five years of German occupation, with food rationing and a severe lack of housing. Often the young couples had to share a house with the husband's parents.

Most of the war brides received warm receptions from their husbands' families, though it could be challenging if the wife's new relatives spoke no English and she spoke no Norwegian. In fact, some of the young women found life in Norway too difficult and returned to Canada. One reporter described the situation:

> Often old-fashioned Norwegian women, finding themselves with foreign daughters-in-law with whom they find communication difficult, are openly critical, stolidly remote. Such small details as the practice of wiping floors with a cloth instead of sweeping them have created great gulfs.[9]

But many of the young Canadian women bloomed in Norway. When Oluf Reed-Olsen's wife, Joan, arrived there, Reed's family instantly made her feel welcome; she enjoyed a very close relationship with them over her lifetime. Joan became fluent in Norwegian and worked as a radio journalist in Norway for many years before she and their young adult son, Jan, moved back to Canada.[10]

Barbara Taylor sailed to Norway on the *Stavangerfjord* to marry Egil Chramer. When Barbara arrived at the family home, Egil's mother greeted her with a big hug; Barbara remembered her mother-in-law as a lovely, kind woman who helped Barbara enjoy a wonderful life in Norway. Rationing made food scarce when Barbara arrived, but Egil's mother regularly bicycled into the country to buy food from farmers to add to the family's diet. This proved especially helpful when Egil and Barbara married, enabling their guests to enjoy a wonderful wedding feast.

Barbara and Egil stayed in Norway for a couple of years

Ole Reistad.

PHOTO: NORWEGIAN ARMED FORCES MUSEUM, OSLO, NORWAY.

after marrying, but then returned to Canada where there were additional opportunities and more plentiful housing.[11]

Packages from Canadians families helped some war brides deal with rationing. Laura Merritt travelled to Norway after the war and married Jakob Anonby in Mysen. Jakob remained with the RNAF post-war and was stationed on a small island off Tønsberg in Oslofjord, from where he maintained cannons on a number of islands. Fish was plentiful in the area; Jakob caught many and taught Laura how to clean and cook them. To help ease the pinch of rationing, each month Laura's mother sent packages from Canada, containing cheese, sugar, tea, and coffee. After a couple of years, Jakob and Laura returned to Canada, bringing their small daughter, Lauretta, along with them.[12]

Many of the war brides kept in close touch, sharing cosmetics, stockings, and other items sent from home, which weren't available in Norway.[13]

A former RNAF airman visited with some of the war brides in Oslo in June 1946, and told *Globe and Mail* that the women's most difficult challenge was a lack of housing, but that the food shortages were easing:

> Some of them have houses for the time being and some have to make [a] go at any kind of an apartment they can get, but they all tell me they like it very much.... Perhaps sometime[s] they are a little homesick, but in Oslo many of the ex-air force personnel are finding new civilian jobs and they still get together and then you ought to hear the English fly! But already many Canadian girls speak Norwegian quite fluently.[14]

LEFT Laura Merritt on the way to Norway to marry her airman, Jakob Anonby. PHOTO: ANONBY FAMILY COLLECTION.

Sale of Vesle Skaugum

In Canada, Vesle Skaugum, with its Norwegian design and beautiful setting, had brought memories of home to many during their air force training days. It bore the stamp of exile, with the Crown Princess's message inscribed on the fireplace in the main log building. The RNAF decided that the property shouldn't be sold to a commercial enterprise, regardless of the price that might be offered. Instead, Vesle Skaugum was sold to the Toronto Kiwanis Club in June 1945, for use as a summer camp for children. The price was $38,000.[15]

Today, Olympia Sports Camp operates a sports-training program in summers for boys and girls at the site of the former Vesle Skaugum. A monument on the grounds commemorates the RNAF. On one side is written:

> In memory of the members of the Royal Norwegian Air Force who gave their lives for freedom in the war 1940–1945. As screams of eagles linger in the air so shall their names ring across the seas, a message of liberty to the hills of home.

Sale of Little Norway

Wilfred Doherty of Gravenhurst purchased the RNAF's property in October 1945 at a cost of $75,000.[16] He planned to open a civilian flying school and air transport service at the former Little Norway. While this project didn't proceed,

TOP The former Vesle Skaugum, now Olympia Sports Camp. Memories of the Royal Norwegian Air Force linger. PHOTOGRAPHER: CANDIS JONES.

BOTTOM The main lodge at Vesle Skaugum, as photographed in 2016. PHOTOGRAPHER: BO GUSTAFSSON.

Crown Prince Haakon and Crown Princess Mette-Marit visit Little Norway Park, Toronto, November 9, 2016.

PHOTOGRAPHER: ANDREA BASTON.

Canada later purchased the property for use as a minimum security prison, Beaver Creek Institution, which opened in 1961. The site remains a medium and minimum security prison today, operated by the Correctional Service of Canada.

Muskoka Airport itself is owned by the District Municipality of Muskoka.

The Norwegian Royal Family

The King honoured some Canadians who had helped Norway in World War II. In 1947, he awarded the Haakon VII Medal of Liberation to eight Ontario doctors for "meritorious services rendered to the Norwegian cause during the war." Dr. H.C.P. Hazlewood, former Superintendent of Muskoka Hospital, received one of these medals posthumously.[17]

King Haakon VII ruled Norway until his death in September 1957, when Crown Prince Olav became King. Crown Princess Märtha died of cancer in 1954, at age 53.

King Olav V reigned for some 34 years, passing away in 1991. He was succeeded by King Harald V, who with his wife, Queen Sonja, reigns today. King Harald and Queen Sonja have two children, Princess Märtha Louise and Crown Prince Haakon Magnus, who is married to Crown Princess Mette-Marit.

Results

Over the course of the war, some 3,323 individuals trained at Little Norway in Toronto and Muskoka.[18] Little Norway sent nearly 2,000 trained air force personnel to the United Kingdom to fight the enemy.

As of June 7, 1945, Norwegian squadrons and personnel serving in the RAF received credit for destroying 252 enemy aircraft, probably destroying 42 others, and damaging 151 more. Five or 6 U-boats were sunk, 4 probably sunk, and 19 badly damaged.

The Norwegian flyers sank 44,000 tons of German transport and damaged another 43,000 tons. In addition, a large number of factories, trains, and enemy planes on the ground were destroyed.

Squadrons

To pay tribute to the wartime successes of RAF Squadrons 331, 332, 333 and 334, the Royal Norwegian Air Force has perpetuated the squadrons' names.

During World War II, Squadrons 331 and 332, together, were credited with 180.5 enemy aircraft destroyed, 25.5 probably destroyed, and 123 damaged. Seventy-two airmen died during operations and 11 became prisoners of war.[22] *Squadrons 331 and 332 are fighter units currently based at Bødo.*

Squadron 330 carried out 655 submarine searches, more than 50 convoy escorts, and 22 rescue missions. It sank 2 U-boats and damaged 4 others, as well as damaging 2 aircraft. Sixty-three personnel died over the course of the war.[23] *The squadron was deactivated after the war but reactivated around 1953. It now flies Sea King helicopters, carrying out search and rescue missions from its base at Sola Air Station.*

Squadron 333 A and B Flights together sank 4 U-boats and damaged 8 more. They destroyed 18 aircraft and damaged 2 others. The squadron also performed 22 special missions to Norway and the Soviet Union. Twenty-eight airmen were lost and 4 became prisoners of war.[24]

On May 26, 1945, Squadron 333 B Flight was designated as Squadron 334. That squadron was transferred to Gardermoen in November 1945. It is still operational within the RNAF, flying military helicopters from Bardufoss Air Station.

Squadron 333 moved to Fornebu Air Base, Norway, in June 1945 and in 1963, to Andøya Air Station. Since then, it has remained active, operating Orion Maritime patrol aircraft in carrying out duties including reconnaissance and search and rescue missions.

Some 309 RNAF members died during the war. Two hundred and ninety-two were killed on active service or were still missing as of June 7, 1945; 15 had died in hospitals; and 2 had been executed as prisoners of war. These statistics include 40 personnel who died in Canada as a result of either accidents or illness. Another 31 airmen were interned as prisoners of war but survived.[19]

The Norwegian merchant marine lost 570 ships. Some 4,000 sailors were killed and about 6,000 ill or injured.[20]

Aces

Sixteen RNAF pilots became aces during the war, meaning they shot down a certain number of enemy aircraft; the number varies but is usually at least five. A practice existed where if two or more pilots claimed the same victory, each was allocated an equal percentage of the win.

Captain Svein Heglund came first, with 16 aircraft destroyed. The other aces were: Lieutenant Colonel Werner Christie, 11; Group Captain Kaj Birksted, 10.5; Captain Helner G.E. Grundt-Spang, 10.33; Major Martin Y. Gran, 9; Captain Marius Eriksen, 9; Captain Nils K. Jørstad, 6.33; Lieutenant Fredrik S. Fearnley, 6.25; Captain Ragner Dogger, 6; Lieutenant Colonel Rolf Arne Berg, 5.75; Lieutenant Helge Sognnæs, 5.5; Captain Bjørn Bjørnstad, 5.5; Captain Ola G. Aanjesen, 5.5; Lieutenant Colonel Helge Mehre, 5; Major Arne Austeen, 5; and Captain Olav Djønne, 5.[21]

Little Norway

Little Norway meant exile to the recruits—living in a country far removed from family and the battles raging in Europe. But Little Norway also meant hope, pride, and the belief that, someday, they would take back their homeland.

Canada helped Norway at a very bleak moment in its history and Canadians can be proud of their country's contributions to Norwegian air training. Although Norway paid the full costs of establishing and operating Little Norway in Toronto and Muskoka, the Toronto Harbour Commissioners provided a rent-free site for the camp and allowed the use of Toronto Island Airport for flight training. Also, more advanced flying lessons were offered to the Norwegian airmen through the British Commonwealth Air Training Plan, administered by Canada. But perhaps the most important gift of all was the friendship which Canadians offered to the young men as they learned to be pilots, navigators and aircraft mechanics.

Once overseas, the Norwegians performed in battle at the highest levels, as RAF Squadrons 331 and 332 demonstrated at Dieppe. The RNAF played its part in liberating Nazi-occupied countries in Europe, and in protecting Canada from enemy incursion. But these successes came at a terrible cost, as many lost their lives.

In the end, the Allies prevailed, World War II ended, and the German occupiers left Norway. Many RNAF members stayed in Canada after the war, or returned here within a few years. They worked, raised families, and contributed to the country's growth and culture. Among those RNAF members who settled in Canada were Jakob Anonby, John Baadsvik, Egil Chramer, Svend Clement, Johan and Elsa Erland, Alf Hansen, Per Conradi Hansen, Rolf Hauge, Per Hvidsten, Rolf Lund, Egon Malmsten, Per Nilsen, Harald Olafson, Ragnar Heiberg Olsen, Kai Pedersen, John Stene, Arne Skretteberg, and Odd Toldnes.

Others went home after the war and made their lives in Norway or Denmark, bringing with them their memories and experiences from Canada. Such individuals include Kaj Birksted, Svein Heglund, Rolf Kolling, Wilhelm Mohr, Per Ole Falkendal Nilsen, Oluf Reed-Olsen, Eigil Stig Stigseth, Finn Thorsager, and Per Waaler.

Canada and Norway are equally enriched by these interwoven knots of friendship, sacrifice, and remembrance.

Epilogue

After the RNAF air-training camps closed, many people searched for a way to honour this remarkable period of Canadian and Norwegian history.

Nova Scotia memorials

In Halifax, a memorial stone was unveiled on November 11, 1998, commemorating members of the Norwegian merchant marine, and the Norwegian Army and Royal Norwegian Navy, who were lost at sea or died in Nova Scotia during the war. Several other monuments honouring the Norwegians' stay in Nova Scotia exist in the province, including those in Lunenburg and Chester.[1]

Granite monument at Little Norway Park, Toronto, November 2016.

PHOTOGRAPHER: ANDREA BASTON.

Ontario memorials

The first Ontario commemoration of the RNAF's air training program took place in September 1976, when Crown Prince Harald, then 39 years old, travelled to Toronto to unveil a plaque and a monument located to the east of the Toronto Island ferry dock at the city's waterfront.

The monument, which is a 1,360-kilogram granite boulder standing on three smaller rocks, came from Lista, Norway. A message of gratitude to Canada for its help and hospitality to the Norwegians during wartime is engraved on the top of the boulder. On one side, a plaque shows two airmen walking on top of the globe, one man standing in Canada and the other in Norway. The design is based on the stamp created at Little Norway around 1942.

Some 150 RNAF veterans and their families flew to Toronto on a specially chartered flight for the monument's unveiling. As the ceremony finished, four Harvard aircraft from the Canadian Warplane Heritage Museum in Hamilton thundered above. The following day, many of the veterans travelled to Muskoka Airport to visit the site of Little Norway.[2] This commemoration proved to be the first of many Canadian reunions for the Norwegians.

In September 1981, a second reunion of RNAF veterans took place in Toronto. Three years later, the Toronto Historical Board and the veterans held a ceremony to unveil a historical plaque at the Toronto Island Airport administration building, commemorating the "Royal Norwegian Air Force: 1940–1945."

The year 1985 represented the 40th anniversary of the closing of the Little Norway camp and a special, week-long reunion took place in June. Some 100 Norwegian veterans returned to Canada to attend the events, which focused on the unveiling of a historical plaque at Muskoka Airport to mark the Norwegians' stay at Little Norway. The plaque was set in a granite stone donated by Gravenhurst's Coon Bros. Sand and Gravel Company.[3]

Little Norway Park

The emergency housing built at the former site of Little Norway in Toronto was torn down in 1958. On November 20, 1987, three hectares of the property became "Little Norway Park." Eighty-four-year-old King Olav returned to Toronto, for the first time since the war ended, to be present for the event.

Mayor Arthur Eggleton greeted the King and the audience of some 800 people, including many dignitaries. In his remarks, King Olav said that because the people of Toronto opened their doors to the air force recruits, Norwegians would always remember the kindness and hospitality of Canadian people.[4] He also re-dedicated the stone boulder monument, which had been moved to Little Norway Park from its original waterfront location.

Albert H. Bacon

Some people felt strongly that Little Norway at Muskoka Airport should receive heritage recognition from the federal government, but no one more so than Al Bacon, a Bracebridge resident.

As a young man, Bacon worked for the Canadian merchant marine on the Great Lakes and Seaway route, serving on the *Kingston*, a steamboat carrying passengers along the St. Lawrence River. Through wartime arrangements made between the Canadian and Norwegian shipping authorities, young mariners, like 16-year-old Bacon, were transferred to sea-going merchant vessels, allowing older sailors to sign up for military duty. Bacon was posted to a Norwegian merchant marine ship for two years. He recalled that the Norwegian sailors treated him wonderfully, as if he was one of their own. In later years, Bacon joined Canada's armed forces.[5]

Individuals had made requests to the federal government since 1963, for Little Norway to be designated as a nationally significant historic site. At that time, the Historic Sites and Monuments Board of Canada had agreed and recommended it be commemorated with a plaque. However, in 1973 the Board revoked its recommendation: Little Norway no longer qualified for designation as a number of its buildings' original features had been lost during the construction of Beaver Creek prison. The board deferred its recommendation.[6]

Around 2000, Norway's Consul General in Toronto, Eivind Hoff, wrote to the Minister of Canadian Heritage, requesting Little Norway's historic importance be commemorated, but met with no success. When he learned about the difficulties Hoff and others had experienced in their efforts to obtain a historic site designation for Little Norway, Al Bacon decided he'd take on the challenge.[7]

Bacon began by seeking support for the designation. As he was a member of the Bracebridge Branch of the Royal Canadian Legion, he spoke with his Legion colleagues and gained their approval. Soon, the District Municipality of Muskoka was behind the idea, too, and public interest was growing.

Then Bacon had a new idea—why not build a memorial at Muskoka Airport to house RNAF artifacts and preserve Little Norway's history?

By early 2001, a joint Canadian-Norwegian committee had begun fundraising, and the District Municipality of Muskoka, which owns and operates the Muskoka Airport, had consented to the creation of a memorial at the airport. Lars Lindgren from Elliot Lake, an exhibit/museum designer who had worked for the Canadian government and private companies, created a conceptual design, storyline, and model for the memorial.[8]

Meanwhile, the Historic Sites and Monuments Board received new guidelines for consideration when recommending properties for designation. Based on these new guidelines, the Board recommended that "Norwegian training in Canada during the Second World War" be designated as an event of national historic significance. In June 2001, the Minister of Canadian Heritage, Sheila Copps, accepted the Board's recommendation and advised Bacon that a plaque would be placed at the airport in Gravenhurst.[9]

On May 8, 2002, King Harald V, who had succeeded his late father as Norway's ruler, unveiled the Board's plaque during a ceremony at Muskoka Airport. He also signed a com-

Albert H. Bacon and Rolf Hauge, at the Canadian Warplane Heritage Museum, Hamilton, following a Remembrance Day service. They are standing in front of a Douglas DC-3. PHOTO: KOOL SHOTS.

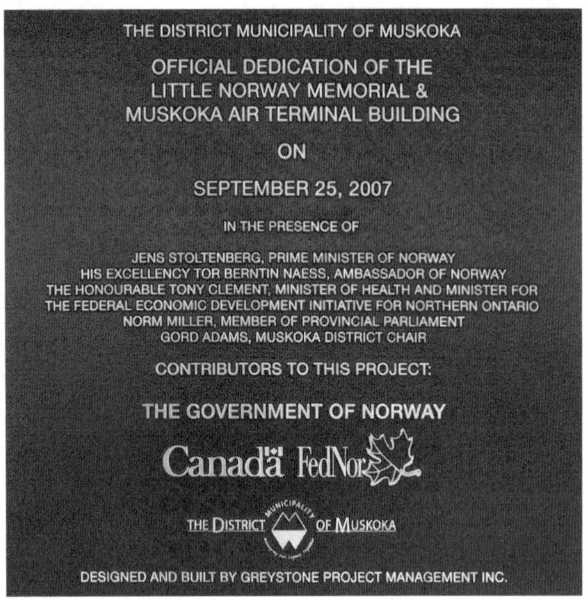

memorative granite cornerstone for the museum, donated by the Norwegian Branch of the Royal Air Force Association.

The crowd numbered around 600, including some 50 Norwegian veterans and their families who had arrived at Muskoka Airport in a Hercules transport plane.[10] A number of the veterans found Canada's formal recognition of the RNAF training to be a moving and emotional event. One man said to Al Bacon, "I thought you'd forgotten us."[11]

There was a final treat: King Harald unveiled an Fairchild training plane marked "*Spirit of Little Norway*." This aircraft had been found in parts, had been carefully restored, and now belonged to the Canadian Warplane Heritage Museum in Hamilton.[12]

In the evening, the Norwegian King and Queen hosted a formal dinner at the Royal York Hotel in Toronto, attended by many prominent Muskoka residents.

Greystone Project Management Inc. of Huntsville was awarded a contract to design and construct a 465-square-metre memorial and cultural centre, which would be connected to the airport building. In 2001, the project's estimated cost was $1.84 million.[13] But fundraising for the project ran into many obstacles, and the designs for the building had to be changed to reflect available financial resources of about $1.1 million. The revised design reduced the memorial's size

TOP Little Norway Memorial and Muskoka Air Terminal building, Muskoka Airport, 2017. PHOTOGRAPHER: CANDIS JONES.

BOTTOM Plaque, marking official opening of Little Norway Memorial and Muskoka Air Terminal building, September 25, 2007. PHOTOGRAPHER: CANDIS JONES.

to some 185 square metres; also, the memorial was integrated into a new, remodelled Muskoka Airport terminal rather than being a separate building.

The *Examiner-Banner* reported that Norway would contribute approximately $350,000 to the cost of the memorial; the District Municipality of Muskoka would provide $333,000; and the federal government would donate up to $430,000 through a program called "FedNor." [14]

Al Bacon continued to work tirelessly for the Little Norway Memorial, through all its phases of development, and when the memorial was completed, he felt "it was a dream come true." [15] Sadly, Al Bacon passed away on December 27, 2014.

Little Norway Memorial opens

The Little Norway Memorial was finally built and it officially opened on September 25, 2007. An audience of about 750 people attended, including some 35 Norwegian veterans. Norway's Prime Minister, Jens Stoltenberg, and Parry Sound-Muskoka's Member of Parliament, Tony Clement, cut the ribbon.

In his remarks, Prime Minister Stoltenberg characterized the Little Norway Memorial as a war memorial, erected to prevent history from forgetting. It also expressed the lasting gratitude of the people of Norway and the Norwegian government to Canada for assistance when Norway needed it most.[16]

The cornerstone, signed by King Harald in 2002, was visible in the memorial's floor. On the walls were exhibits, photographs, and information panels telling Little Norway's story, designed by Lars Lindgren, who also served as the memorial's curator.[17] A beautiful, cabin-like room, with log walls and a

John and Inger Stene at opening of Little Norway Memorial. John is standing beside a mannequin wearing his old uniform; he later tried on the old uniform and was delighted to see that it still fit perfectly. PHOTO: INGER STENE COLLECTION.

fireplace, offered a visitor a comfortable spot in which to pore over information and pictures borrowed from bookcases.

John Stene was among the RNAF veterans present, along with his wife, the former Resistance fighter Inger (Dietrichson) Stene. John's old uniform was being worn by a mannequin in a display case; it bore his medals, including the Distinguished Flying Cross with a bar, signifying that Britain had twice awarded him that medal for acts of valour during active operations. While at the memorial, John decided to try on his old uniform and was delighted to learn that it still fit perfectly.[18]

Wall plaques listing RNAF wartime casualties were added to the Little Norway Memorial on May 31, 2012. Lieutenant General Wilhelm Mohr, aged 95, came from Norway with others for the ceremony. He said, "I can only give all the credit that's at all possible to the Canadian people, the air force and to the local people of this area for the contribution and the help we got."[19]

Today, as the Little Norway Memorial exemplifies, the Royal Norwegian Air Force air training camps in Toronto and Muskoka remain a significant, respected part of the history of both Norway and Canada.

Royal Norwegian Air Force pilots beside row of Fairchild PT-19 training planes, Toronto Island Airport, 1941.

PHOTO: LITTLE NORWAY MEMORIAL.

Notes

Introduction

1 "Training Base of Norwegians Declared Open," *Globe and Mail*, November 11, 1940.

2 This book does not attempt to recount the full history of the German invasion of Norway and subsequent battles, but only to touch on some of the crucial events.

Chapter 1

1 International Military Tribunal, *Trial of the Major War Criminals before the International Military Tribunal, Nuremberg, 14 November 1945 – 1 October 1946*, vol. 3 (Nuremberg, Germany: International Military Tribunal, 1947), 265. There were also other reasons for German naval leaders to argue that Norway should be invaded and occupied. For instance, an occupation would preclude the British from setting up a mine barrage across Norwegian waters to restrict the movement of German ships, as had occurred in World War I.

2 International Military Tribunal, *Trial of the Major War Criminals before the International Military Tribunal, Nuremberg, 14 November 1945 – 1 October 1946*, vol. 14 (Nuremberg, Germany: International Military Tribunal, 1948), 90, 91; Germany: Kriegsmarine Oberkommando, *Fuehrer Conferences on Matters Dealing with the German Navy 1940*, vol. 1 (Washington, D.C.: U.S. Navy Department, 1947), 14.

3 "Could Help Russ Problem by Recapturing Norway," *Toronto Daily Star*, August 13, 1942. Comments of General C.G. Fleischer.

4 International Military Tribunal, *Trial of the Major War Criminals before the International Military Tribunal*, vol. 3, 274.

5 United Kingdom, HC Deb May 2 1940, vol 360, col 907.

6 "Nazi Bullies on *Altmark* Kicked 13 Year-Old Lad," *Toronto Daily Star*, September 10, 1941.

7 "Canuck Pilot Promises *Altmark* Picture Feast," *Toronto Daily Star*, March 27, 1940; "Men Went Mad on *Altmark* Says Man Who Was There," *Toronto Daily Star*, June 6, 1940; Henrik O. Lunde, *Hitler's Pre-Emptive War: The Battle for Norway, 1940* (Havertown, Pennsylvania: Casemate, 2010), 26, 27.

8 William L. Shirer, *The Rise and Fall of the Third Reich: A History of Nazi Germany* (New York, New York: Simon & Schuster Paperbacks, 1959), 679, states that four died; Lunde, *Hitler's Pre-Emptive War*, 31, writes: "five (seven according to some sources) dead."

9 "Nazi Bullies on *Altmark* Kicked 13 Year-Old Lad," *Toronto Daily Star*, September 10, 1941.

10 Germany, "Directive for Weserübung," *Documents on German Foreign Policy 1918–1945*, series D, vol. 8 (Washington, D.C.: U.S. Government Printing Office, 1954), 831–833.

11 Ibid.; International Military Tribunal, *Nazi Conspiracy and Aggression*, vol. 6 (Washington, D.C.: U.S. Government Printing Office, 1946), 302.

12 International Military Tribunal, *Trial of the Major War Criminals before the International Military Tribunal*, vol. 3, 131; Lunde, *Hitler's Pre-Emptive War*, 74–77.

13 "Allies Mine Norse Waters Glad to Have Nazis Object," *Toronto Daily Star*, April 8, 1940; "Allies Mine Norse Waters Cripple All German Trade Hope to Bottle Nazi Fleet," *Evening Telegram*, April 8, 1940.

14 Numbers of aircraft possessed by the services vary from account to account, but these numbers are derived from: H.W. Looseley, "History of the Royal Norwegian Air Force in Canada," F, G, History, Toronto, 1945, L Box, Class 385.4 (481) Norwegian Armed Forces Museum Library, Oslo, Norway.

Chapter 2

1 "Reich Ship is Sunk," *New York Times*, April 9, 1940; "Soldiers Seen Leaping Off Ship that Ignored Sub's Orders to Stop," *Toronto Daily Star*, April 8, 1940; "British Ship Goes Down Sinking Sub," *Evening Telegram*, April 8, 1940; Shirer, *The Rise and Fall of the Third Reich*, 695.

2 W.R. Plewman, "The War Reviewed," *Toronto Daily Star*, April 8, 1940; "Heavy German Warships Moved to Norse Coast Landing Forces Stand By," *Evening Telegram*, April 8, 1940.

3 Lunde, *Hitler's Pre-Emptive War*, 93.

4 International Military Tribunal, *Trial of the Major War Criminals before the International Military Tribunal*, vol. 3, 283, 284; "Goebbels Explains Move," *New York Times*, April 9, 1940.

5 Forsvarets Museer, "9 April 1940," Forsvaretsmuseer.no, http://forsvaretsmuseer.no/Oscarsborg/Festningens-historie/9.april-1940 (Accessed March 7, 2017); World War II Today, "Norwegian Shore Batteries Sink the Blucher," ww2today.com, http://ww2today.com/norwegian-shore-batteries-sink-the-blucher (Accessed March 7, 2017).

6 "Off to Invade England Nazi Idea Till Oslo Near," *Toronto Daily Star*, April 18, 1940.

7 Lunde, *Hitler's Pre-Emptive War*, 220, cites "about 1,000"; Shirer, *The Rise and Fall of the Third Reich*, 703, refers to 1,600.

8 Tor Idar Larsen and Finn Thorsager, *Viking Spitfire: The Story of Finn Thorsager* (United Kingdom: Fonthill Media, 2012), 37.

9 Ibid., 42–47.

10 Ibid.; Spitfirepilots, "Per Waaler—Gloster Gladiator Pilot, " Spitfirepilots.com, http://www.spitfirepilots.com/per-waaler-gloster-gladiator-pilot/ (Accessed March 7, 2017); Håkan Gustavsson, "Biplane Fighter Pilots Norway: Sergeant Per Waaler," Håkan's Aviation Page, http://surfcity.kund.dalnet.se/norway_waaler.htm (Accessed March 7, 2017); Håkan Gustavsson, "Biplane Fighter Pilots Norway: Major Finn Thorsager," Håkan's Aviation Page, http://surfcity.kund.dalnet.se/norway_thorsagen.htm (Accessed March 7, 2017).

11 Lunde, *Hitler's Pre-Emptive War*, 220, indicates the number was five; Olav Myre, United Press Staff Correspondent, in an untitled story, *Toronto Daily Star*, April 9, 1940, is quoted as saying: "I believe I saw four German planes crash."; Wall Text, "Invasion of Norway, April 9, 1940," Little Norway Memorial, 2007, Muskoka Airport, Gravenhurst, Ontario, states that six German aircraft were destroyed.

12 Kjell Arne Reistad, Personal Interview with Author, Oslo, Norway, August 24, 2015; Kai Pedersen, "Little Norway May She Always Be Free," *The Forester*, August 19, 1981; Kai Pedersen, *Viking Ships Had No Lifeboats and Other Stories* (Burk's Falls, Ontario: Arrow Printing and Publishing, 1995), 29.

13 Peter C. Rhodes, "2,000 Germans Kill 542 Norse to Take Narvik," *Toronto Daily Star*, April 10, 1940; "Helpless Norse Sailors Machine-Gunned in Water," *Toronto Daily Star*, April 24, 1940; Shirer, *The Rise and Fall of the Third Reich*, 701; Lunde, *Hitler's Pre-Emptive War*, 162–164. There are differing accounts as to whether the *Wilhelm Heidkamp* fired on the *Eidsvold* as soon as the launch bearing the German negotiator was sufficiently far away from the Norwegian ship. Shirer suggests that this was the case. However, Lunde writes that the German officer had returned to the *Wilhelm Heidkamp*, and that the *Eidsvold* was heading towards the German destroyer before the latter fired its torpedoes.

14 Lunde, *Hitler's Pre-Emptive War*, 165.

15 "Every Enemy Warship in Norwegian Waters Bottled Up by Allies," *Toronto Daily Star*, April 10, 1940. A British submarine sank *Karlsruhe* as it headed home to Germany on the night of April 9–10, 1940.

16 "We Hit Germans' Ship Explosion Was Terrific," *Toronto Daily Star*, April 12, 1940; "Watched R.A.F. Bombers Sink Nazi Cruiser," *Toronto Daily Star*, May 3, 1940; Shirer, *The Rise and Fall of the Third Reich*, 701; M.J. Pearce and R. Porter,Introduction, *Fight for the Fjords: The Battle for Norway 1940*, ed. by Charlotte Carey (Plymouth, England: University of Plymouth Press, 2012), 12. Some 15 British Skua dive-bombers flew to Norway and sank *Königsberg* in Bergen harbour early in the morning of April 10. The *Königsberg* is said to have been the first major warship sunk by air attack during World War II.

17 Dag Henriksen, "Biography of Wilhelm Mohr," *Wilhelm Mohr on World War II*, comp. by Dag Henriksen (Trondheim, Norway: Tapir Academic Press, 2009), 13.

18 Shirer, *The Rise and Fall of the Third Reich*, 702.

19 International Military Tribunal, *Trial of the Major War Criminals before the International Military Tribunal*, vol. 3, 285–287; International Military Tribunal, *Nazi Conspiracy and Aggression*, vol. 6, 304–307; Svend

Carstensen, "Copenhagen Taken," *New York Times*, April 9, 1940; "Norse Giving In, Berlin says, but Only Inch at a Time," *Toronto Daily Star*, April 10, 1940; "Stunned Danes Awake to Find Armed Nazis Posted Throughout City," *Evening Telegram*, April 10, 1940; Shirer, *The Rise and Fall of the Third Reich*, 699.

Chapter 3

1 Knut Fossum, Personal Interview with Author, Oslo, Norway, August 25, 2015; Randi Hvatum, Telephone Interview with Author, May 16, 2014; "Oslo Folk Flee in Droves Fearing British-Nazi Fight," *Toronto Daily Star*, April 11, 1940.

2 Gunhild Ramm Reistad, Personal Interview with Author, Oslo, Norway, August 24, 2015.

3 John Magne Stene, "The Distinguished Flying Cross and How It Was Won," Autobiography, Toronto, 1995, Inger Stene Collection, 6, 7; Henriksen, "Biography of Wilhelm Mohr," *Wilhelm Mohr on World War II*, 14; Wilhelm Mohr, "9 April—From the Past to the Future," *Wilhelm Mohr on World War II*, comp. by Dag Henriksen (Trondheim, Norway: Tapir Academic Press, 2009), 103, 104.

4 Robert Pearson, *Gold Run: The Rescue of Norway's Gold Bullion from the Nazis, April 1940* (Oxford, England: Casemate, 2015), 44; Lunde, *Hitler's Pre-Emptive War*, 226; Shirer, *The Rise and Fall of the Third Reich*, 698.

5 United Kingdom, HC Deb April 9 1940, vol 359, col 509; "Powerful Units of Navy and Huge Allied Air Force Join Battle Beside Norse," *Evening Telegram*, April 9, 1940.

6 Robert Pearson, *Gold Run*, 55, 241. While descriptions of the value of the gold rescued differ from story to story, this currency conversion was based on: Norges Bank, "Historical Exchange Rate Data, 1819–2003," norges-bank.no, http://www.norges-bank.no/en/Statistics/Historical-monetary-statistics/Historical-exchange-rates/ (Accessed March 9, 2017). Note: Descriptions of the gold's escape from the Nazis in this book are based on Pearson, *Gold Run*; Kurt Singer, *Duel for the Northland: The War of Enemy Agents in Scandinavia* (New York, New York: Robert M. McBride & Company, 1943), 82–90; "Gold Out by Back Door as Nazi Entered Front," *Toronto Daily Star*, January 22, 1941.

7 "Gold Out by Back Door as Nazi Entered Front," *Toronto Daily Star*, January 22, 1941.

8 "Norse Royalty Safe After Long Voyage," *Evening Telegram*, August 29, 1940.

9 Arthur Menken, "Norwegian College Boys Thumb Way Home to War," *Toronto Daily Star*, April 15, 1940.

10 This became known as the "Elverum Authorization."

11 "Nazi Bombs Rain All Night Swedes Fear Next Thrust," *Toronto Daily Star*, April 12, 1940; Lunde, *Hitler's Pre-Emptive War*, 227, 228; Shirer, *The Rise and Fall of the Third Reich*, 704.

12 "'Haven't Taken Off Boots Since Tuesday'—Haakon 'Civilization At An End,'" *Toronto Daily Star*, April 13, 1940; "Haakon Hounded by Nazis Refuses to Leave Country," *Evening Telegram*, April 13, 1940.

13 Rolf Kristian Hauge, Personal Interview with Author, Collingwood, Ontario, May 6, 2015.

Chapter 4

1 Lunde, *Hitler's Pre-Emptive War*, 265.

2 Ibid., 270.

3 United Kingdom, HC Deb May 7 1940, vol 360, cols 1076, 1077; Lunde, *Hitler's Pre-Emptive War*, 235, 236.

4 Lunde, *Hitler's Pre-Emptive War*, 314.

5 United Kingdom, HC Deb May 2 1940, vol 360, col 909; United Kingdom, HC Deb May 8 1940, vol 360, cols 1353, 1354; "Report British Advancing to Major Clash with Foe in North Trondheim Zone," *Evening Telegram*, April 18, 1940.

6 United Kingdom, HC Deb May 2 1940, vol 360, col 910.

7 Henriksen, "Biography of Wilhelm Mohr," *Wilhelm Mohr on World War II*, 14.

8 Ibid.; United Kingdom, HL Deb May 8, 1940, vol 116, cols 296, 297; "Cheers Echo in Commons as Challenge of Labour 'Welcomed' by Premier," *Evening Telegram*, May 8, 1940; "A.A. Guns on Way to Norway Sunk by Sub, Hankey Says," *Toronto Daily Star*, May 8, 1940; World War II Today, "263 Squadron Land at Lake Lesjaskog," ww2today.com., http://ww2today.com/24th-april-1940-263-squadron-land-at-lake-lesjaskog (Accessed March 9, 2017).

9 C.B. Pyper, "Brilliant Norway Action Shows British Better Men Than

Nazis Ever Will Be," *Evening Telegram*, May 7, 1940; "Norwegian Crews Drove Trains Right up to Front to Get Them Returning British Soldiers Say," *Toronto Daily Star*, May 7, 1940.

10 Roy Jenkins, *Churchill: A Biography* (New York, New York: Penguin Group, 2002), 575–588.

11 A collection of Nordahl Grieg's war poems called *Friheten* (Freedom) remains popular in Norway today. Grieg died on December 2, 1943, when a Lancaster bomber, in which he was flying, was shot down during an air raid over Berlin.

12 Robert Pearson, *Gold Run*, 121–125. The photograph was taken by Norwegian Foreign Affairs Press Office photographer Per Bratland.

13 Knut Larsson, "RAF Nordic Casualties of Crews and Losses of Aircraft in World War 2," luftwaffe.no, http://www.luftwaffe.no/Table2.htm (Accessed March 9, 2017). In 1944, Niels Christian Evensen was captain of a Lancaster bomber heading for Munich when his plane was shot down near Dieppe and he was killed.

14 This currency conversion was based on: Norges Bank, "Historical Exchange Rate Data, 1819–2003," norges-bank.no, http://www.norges-bank.no/globalassets/upload/import/publikasjoner/skriftserie/35/chapter7.pdf (Accessed May 6, 2017).

15 Stene, "The Distinguished Flying Cross and How It Was Won," 7, 8.

Chapter 5

1 Lunde, *Hitler's Pre-Emptive War*, 491–499.

2 Ibid., 515, 516.

3 Ibid., 523, 524.

4 Ibid., 534–538.

5 Norway, *HM King Olav V of Norway*, Pamphlet, n.d., Rita Hiorth Collection.

6 "End of War in Norway Leaves Nazi Sea Route Open," *Evening Telegram*, June 10, 1940; Johs. Andenæs, Olav Riste, Magne Skodvin, *Norway and the Second World War* (Oslo, Norway: Johan Grundt Tanum Forlag, 1966), 93.

7 Norway, "The History of 'Little Norway' in Canada," History, Ottawa, 1987, PortsToronto Archives, Toronto, Ontario.

8 Wilhelm Mohr, "The Contribution of the Norwegian Air Forces," *Wilhelm Mohr on World War II*, comp. by Dag Henriksen (Trondheim, Norway: Tapir Academic Press, 2009), 32.

9 "Buchenwald Victim to be Stalin Guest," *Toronto Daily Star*, June 4, 1945. Major General Otto Ruge was released from Buchenwald Concentration Camp when the Red Army liberated the camp in 1945.

10 Aaron, Robert, "Hoard of Gold Coins Hidden During War Returned to Norway," *Toronto Star*, January 23, 1988. Robert Pearson in *Gold Run*, 226, notes that King Olav personally accompanied the gold as it travelled from the U.S. to Norway by SAS airline on December 1, 1987.

11 Stene, "The Distinguished Flying Cross and How It Was Won," 8, 9.

Chapter 6

1 Norway, "Norway and Canada during World War 2," Royal Norwegian Embassy in Ottawa, https://www.norway.no/en/canada/norway-canada/historical-ties/#NorwayandCanadaduringWorldWar2 (Accessed March 9, 2017); Andenaes et al., *Norway and the Second World War*, 96.

2 Norway, "Norway and Canada during World War 2," Royal Norwegian Embassy in Ottawa, https://www.norway.no/en/canada/norway-canada/historical-ties/#NorwayandCanadaduringWorldWar2 (Accessed March 9, 2017).

3 Ibid.; "The Norwegians Were Once Great Whale Fishers," *Toronto Daily Star*, January 29, 1942. During the war, many of the whaling ships were converted to naval warships and hunted submarines.

4 Pedersen, *Viking Ships Had No Lifeboats and Other Stories*, 182.

5 Wilhelm Mohr, "The Contribution of the Norwegian Air Forces," *Wilhelm Mohr on World War II*, 34.

6 Ibid., 40, 72–77.

7 Canada, "The British Commonwealth Air Training Plan," Veterans Affairs Canada, http://www.veterans.gc.ca/eng/remembrance/history/second-world-war/british-commonwealth-air-training-plan (Accessed March 10, 2017); Juno Beach Centre, "British Commonwealth Air Training Plan," junobeach.org, https://www.junobeach.org/canada-in-wwii/articles/british-commonwealth-air-training-plan/ (Accessed March 10, 2017).

8 Canada, Parliament, House of Commons, *Debates*, 19th Parl., 1st sess., vol. 1, May 20, 1940, 43. Costs of the BCATP were forecast to total about

$600 million and each of the four signatory countries would contribute an assigned percentage. Because it was short of foreign currency, the United Kingdom agreed to supply most of the aircraft instead of making a cash payment. Canada's responsibility amounted to around $350 million, a substantial sum for the country to bear. Each country was also allocated a percentage of students to be trained under the BCATP and for Canada, this amounted to some 80 per cent of the recruits.

9 Wall Text, *Commonwealth Air Training Plan*, Canadian Warplane Heritage Museum, Hamilton, Ontario.

10 Canada, Parliament, House of Commons, *Debates*, 19th Parl., 1st sess., vol. 1, May 27, 1940, 253, 256.

11 Ibid., 255; "Canada Airmen are Preferred as Instructors," *Evening Telegram*, May 7, 1940.

12 Michael Skretteberg, Personal Interview with Author, Claremont, Ontario, December 8, 2014.

13 John Clement, Email Interview with Author, June 3, 2014; Ships of the East Atlantic Line on Postcards, *Europa*, mariners-l,co.uk, http://www.mariners-l.co.uk/eac/eac1930-1939.htm (Accessed March 10, 2017). In May 1941, the *Europa* was destroyed by Luftwaffe bombs while in Liverpool, England. Fortunately, only one person was slightly injured in the attacks.

Chapter 7

1 Looseley, "History of the Royal Norwegian Air Force in Canada," 1.

2 "Norse Fliers Swim, Walk, Row to Rejoin Legion Here," *Toronto Daily Star*, August 6, 1940.

3 Hj. Riiser-Larsen and B.F. Motzfeldt to W.J.E. Johnston, Letter, July 1, 1940, RG 3/3, Box 33, PortsToronto Archives, Toronto, Ontario.

4 Looseley, "History of the Royal Norwegian Air Force in Canada," 1.

5 "Here to Pick Up Warplanes 100 of Them for Norway," *Toronto Daily Star*, April 26, 1940.

6 D.G. Joy to Fred Scandrett, Letter, August 15, 1940, RG 3/3, Box 33, PortsToronto Archives, Toronto, Ontario; "Permit Norwegian to Train Airmen," *Evening Telegram*, July 5, 1940.

7 J. A. Wilson to General W. Steffens, Letter, August 13, 1940, RG 3/3, Box 33, PortsToronto Archives, Toronto, Ontario.

8 Hj. Riiser-Larsen and B.F. Motzfeldt to W.J.E. Johnston, Letter, July 1, 1940, RG 3/3, Box 33, PortsToronto Archives, Toronto, Ontario; E. L. Cousins to Chairman and Members, The Toronto Harbour Commission, Letter, July 4, 1940, RG 3/3, Box 33, PortsToronto Archives, Toronto, Ontario.

9 "125 Viking War Aces to Roam Sky's Lanes over Island Airport," *Toronto Daily Star*, July 4, 1940.

10 Per Conradi Hansen, *Little Norway: A Message of Liberty to the Hills of Home* (Oslo, Norway: The Military Historical Foundation of Eastern Norway, 1985), 21; Scott McClellan, "Recalling the Spirit of Little Norway, Part I: The Escape," *Muskoka Sun*, June 6, 1985.

11 Daniel Steen to E.L. Cousins, Letter, July 11, 1940, RG 3/3, Box 33, PortsToronto Archives, Toronto, Ontario.

12 His Majesty King Haakon of Norway and the Norwegian Government and The Toronto Harbour Commissioners, Agreement, September 13, 1940, RG 3/3, Box 33, PortsToronto Archives, Toronto, Ontario.

13 "Norway Plans Training Pilots at Local Field," *Evening Telegram*, July 4, 1940.

14 "Norwegian Pilots to Train Here for Further Combat," *Toronto Daily Star*, July 20, 1940.

15 Hansen, *Little Norway*, 25.

16 Cato Guhnfeldt, *Spitfire Saga*, vol. 1 (Oslo, Norway: Wings Forlag, 2009), 32.

17 Harold North, "Beyond the Seas," Manuscript, Gravenhurst, n.d., 295, 296, 303–308, Olafson Family Collection; "Ship Which Survived Two Wars Will House Norse Fliers Here," *Toronto Daily Star*, September 6, 1940; Warsailors, D/S *Iris*, warsailors.com., http://www.warsailors.com/singleships/iris.html (Accessed March 10, 2017); Warsailors, D/S *Lyra*, warsailors.com., http://www.warsailors.com/singleships/lyra.html (Accessed March 10, 2017).

Chapter 8

1 Wall Text, "Little Norway," Little Norway Memorial, 2007, Muskoka Airport, Gravenhurst, Ontario; F. J. Hatch, "The Spirit of Little Norway," *Canada's Wings: High Flight*, vol. 1, no. 5, April 1981, 173.

2 Gunhild Tegen, "Little Norway," *Wings for Norway*, 1942, 17, Karen

Greavette Collection.

3 F.R. Scandrett to V.I. Smart, Letter, August 13, 1940, RG 3/3, Box 33, PortsToronto Archives, Toronto, Ontario.

4 Bjarne Øen and O.B. Stene to Department of Transport, Letter, October 3, 1940, RG 3/3, Box 33, PortsToronto Archives, Toronto, Ontario.

5 Plaque, "The Royal Norwegian Air Force: 1940–45," Toronto Historical Board, 1984, Toronto, Ontario; Looseley, "History of the Royal Norwegian Air Force in Canada," 9; Hansen, *Little Norway*, 26, 28.

6 Harald Jensen, *The Adrenalin Years* (Don Mills, Ontario: Frances M. Jensen, 1992), 65.

7 "Norse Camp Inaugurated as Flyers Continue Fight to Lift Yoke from Nation," *Evening Telegram*, November 11, 1940.

8 Sigrid Schultz, "The Spirit of Little Norway," *Chicago Daily Tribune*, August 16, 1942. Hansen, *Little Norway*, 45, indicates that Little Norway was originally constructed at a cost of around $300,000, including the expense of building a hangar at Toronto Island Airport.

9 "Between You and Me," *Evening Telegram*, November 11, 1940.

10 "Training Base of Norwegians Declared Open," *Globe and Mail*, November 11, 1940; "Vikings' Flag Flies Bravely Norse Warplanes Overhead," *Toronto Daily Star*, November 11, 1940; "Norse Camp Inaugurated as Flyers Continue Fight to Lift Yoke from Nation," *Evening Telegram*, November 11, 1940.

11 "Over the Teacups," *Toronto Daily Star*, November 11, 1940.

12 "Between You and Me," *Evening Telegram*, November 11, 1940.

13 Neil Etienne, "Connection to Norway Honoured," *Gravenhurst Banner*, June 6, 2012.

14 Scott McClellan, "Recalling the Spirit of Little Norway, Part I: The Escape," *Muskoka Sun*, June 6, 1985.

15 S.J. Reginald Saunders, *Little Norway in Pictures: R.N.A.F. in Canada* (Toronto, Ontario: Norwegian Air Force in Exile, 1943), n.p.; Hansen, *Little Norway*, 35.

16 Tegen, "Little Norway," *Wings for Norway*, 16.

17 Michael Skretteberg, Personal Interview with Author, Claremont, Ontario, December 8, 2014.

18 Looseley, "History of the Royal Norwegian Air Force in Canada," 48.

19 "Picked 7 War Victors U.S. Flier Joins Norse," *Toronto Daily Star*, July 7, 1941.

20 "Norse Airman Now Convinced That Sign Meant What It Said," *Toronto Daily Star*, September 28, 1940; Hj. Riiser-Larsen and A.W. Hagtvedt to W.J.E. Johnston, Letter, September 26, 1940, RG 3/3, Box 33, PortsToronto Archives, Toronto, Ontario; W.J.E. Johnston to E.L. Cousins, Letter, September 27, 1940, RG 3/3, Box 33, PortsToronto Archives, Toronto, Ontario.

21 Rolf Kolling, Personal Interview with Author, Oslo, Norway, August 25, 2015; Warsailors, M/S *Hav*, Warsailors.com, http://www.warsailors.com/singleships/hav.html (Accessed March 12, 2017).

22 Hansen, *Little Norway*, 35, 64, 117, 118; Looseley, "History of the Royal Norwegian Air Force in Canada," 51; Canada, "The British Commonwealth Air Training Plan," Veterans Affairs Canada, http://www.veterans.gc.ca/eng/remembrance/history/second-world-war/british-commonwealth-air-training-plan (Accessed March 12, 2017); Juno Beach Centre, "British Commonwealth Air Training Plan," junobeach.org, http://www.junobeach.org/canada-in-wwii/articles/british-commonwealth-air-training-plan/ (Accessed March 12, 2017).

Chapter 9

1 Gunnar Sønsteby, *Report from #24* (Fort Lee, New Jersey: Barricade Books Inc., 1999), 28.

2 Lieutenant "Nils Olsen," "Norse Fugitives' Ship Hit Downed Two Raider Planes," *Toronto Daily Star*, November 13, 1941.

3 Barbara Chramer, Personal Interview with Author, Caledon, Ontario, September 18, 2015; J. Peter Hvidsten, "Per Hvidsten: An Amazing Journey, Norway to Canada," Biography, Port Perry, Ontario, 2012, 8–20, J. Peter Hvidsten Collection.

4 Oluf Reed-Olsen, *Two Eggs on My Plate* (London, England: George Allen & Unwin Ltd., 1952), 46.

5 Ibid., 11–91; Patrick Milsom, "Norse 'Ulysses' Outwits Nazis—Now He Can Fight Back," *Toronto Daily Star*, November 15, 1940.

6 David Howarth, *The Shetland Bus* (Guilford, Connecticut: The Lyons Press, 1951), 5, 6, 221, 224, 229.

7 John Stene, "The Distinguished Flying Cross and How It Was Won," 9–12.

Chapter 10

1 Wall Text, "Hitler Drops Quisling," Norway's Resistance Museum, Oslo, Norway.
2 Olav Riste and Berit Nökleby, *Norway 1940–45: The Resistance Movement*, 3rd ed. (Oslo, Norway: AIT Otta AS, 2004), 10, 11.
3 International Military Tribunal, *Trial of the Major War Criminals before the International Military Tribunal*, vol. 3, 287.
4 Mark Mazower, *Hitler's Empire: How the Nazis Ruled Europe* (London, England: Penguin Books, 2008), 104.
5 Riste and Nökleby, *Norway 1940–45: The Resistance Movement*, 30.
6 Ibid., 11–13; Andenæs et al., *Norway and the Second World War*, 94, 95.
7 Rolf Hauge, Personal Interview with Author, Collingwood, Ontario, May 6, 2015.

Chapter 11

1 Jensen, *The Adrenalin Years*, 65.
2 "Flier Still Unconscious in First Norse Air Crash," *Toronto Daily Star*, October 24, 1940.
3 "2 Norse Fliers Hurt in Crash," *Globe and Mail*, October 24, 1940; "Flier Still Unconscious in First Norse Air Crash," *Toronto Daily Star*, October 24, 1940; "Norse Victims of Air Smash Badly Injured," *Evening Telegram*, October 24, 1940; "Unconscious Eight Days," *Toronto Daily Star*, October 31, 1940.
4 "Flying to Dead Wife, Norwegian Airman Crashes," *Toronto Daily Star*, December 27, 1940; "Pilot Lands Safely but Plane Hits Tree," *Globe and Mail*, December 27, 1940; "Two Norwegian Fliers Unhurt in Plane Crash," *Globe and Mail*, December 27, 1940.
5 Looseley, "History of the Royal Norwegian Air Force in Canada," 41.
6 "With Canada's Airmen," *Toronto Daily Star*, October 15, 1940.
7 "Norwegian Army Airmen Train at Island Airport to Strike Another Blow for Homeland," *Toronto Daily Star*, October 24, 1940.
8 Cyril Fry, Personal Interview with Author, Gravenhurst, Ontario, March 29, 2016. Email interview with Author, March 4, 2017.
9 Percy T. Cole, "Fighters Technical Winners as Norse 'Bomb' Toronto But Island Port 'In Ruins'," *Evening Telegram*, March 10, 1941.
10 Randi Hvatum, Telephone Interview with Author, May 16, 2014.
11 Norway, "The History of 'Little Norway' in Canada," 4; Looseley, "History of the Royal Norwegian Air Force in Canada," 1, 17.
12 Odd Toldnes, Audio Tape interview with Gravenhurst Archives, in "World War II Interviews: Muskoka Pioneers," 1989, Tape 97–18, Gravenhurst Archives, Gravenhurst, Ontario; Karen Greavette, Personal Interview with Author, Gravenhurst, Ontario, May 14, 2014.

Chapter 12

1 "Bay Boats Ground to Bits Like Toys in Ripping Gale," *Toronto Daily Star*, September 26, 1941.
2 "Over the Teacups," *Toronto Daily Star*, October 3, 1940.
3 Scott McClellan, "Recalling the Spirit of Little Norway, Part I: The Escape," *Muskoka Sun*, June 6, 1985.
4 "Mary Frances Loosemore Norwegian Airman's Bride," *Toronto Daily Star*, December 26, 1940; "Escaped from Nazis, Norwegian Weds Toronto Girl," *Toronto Daily Star*, December 27, 1940.
5 "Norse Flying Student to Marry Toronto Girl," *Globe and Mail*, January 4, 1941.
6 "Mary Falla is Married to Lieut. H. Olafson," *Globe and Mail*, February 18, 1942.
7 Michael Skretteberg, Personal Interview with Author, Claremont, Ontario, December 8, 2014.
8 Hansen, *Little Norway*, 42.
9 W.J.E. Johnston to E.L. Cousins, Letter, June 24, 1941, RG 3/3, Box 33, PortsToronto Archives, Toronto, Ontario.
10 O. Hughes, Memorandum, July 4, 1941, RG 3/3, Box 33, PortsToronto Archives, Toronto, Ontario.
11 W.J.E. Johnston to E.L. Cousins, Letter, October 29, 1941, RG 3/3, Box 33, PortsToronto Archives, Toronto, Ontario.
12 Looseley, "History of the Royal Norwegian Air Force in Canada," 34, 35.
13 Rolf Kolling, Personal Interview with Author, Oslo, Norway, August 25, 2015.
14 "Norwegians Attend Supper," *Evening Telegram*, December 23, 1940; "Norwegians Guests of Kiwanis Women," *Toronto Daily Star*, December 23, 1940; "Canadian and Norwegian Airmen Get Turkey Feast," *Evening*

Telegram, December 24, 1940; "Airmen Get Turkey Feed at Howard Park Church," *Toronto Daily Star*, December 24, 1940.

15 "Soldier for Dinner Plan Over the Top in Toronto," *Toronto Daily Star*, December 27, 1940; "Soldiers and New Year Parties," *Toronto Daily Star*, January 2, 1941.

16 "Their Old Yule Customs are Revived in Toronto by Norse Fighting Men," *Globe and Mail*, December 25, 1940.

17 "Norse Airmen to Share Yule Gifts with Sailors," *Globe and Mail*, December 25, 1941.

18 Hansen, *Little Norway*, 45.

19 Rolf Kolling, Telephone Interview with the Author, January 16, 2015.

20 Elaine Hooper, Personal Interview with Author, Toronto, Ontario, February 22, 2016.

21 Bill Whiteside and Susan Whiteside Davis, Personal Interview with Author, Toronto, Ontario, July 22, 2015.

Chapter 13

1 Jensen, *The Adrenalin Years*, 69.

2 "Active Service Canteen is Well Patronized," *Evening Telegram*, January 29, 1941; Citizens Committee for Recreation of Troops in Training, Active Service Canteen Toronto 1939–1945 (Toronto, Ontario: Bryant Press Ltd., 1945), n.p.

3 "Have Danced '1,000 Miles' with Allied Servicemen," *Toronto Daily Star*, January 24, 1944.

4 Ibid.; Jensen, *The Adrenalin Years*, 68.

5 Rolf Kolling, Telephone Interview with Author, January 18, 2015.

6 Laura Anonby and Lauretta Anonby, Email Interview with Author, July 29, 2015.

7 "Lyric Male Choir Host to Norwegian Singers," *Globe and Mail*, April 8, 1942; "Norwegian Airmen-Singers to Join Coliseum Chorus," *Toronto Daily Star*, April 14, 1942; "Grace Notes," *Globe and Mail*, May 2, 1942; "Men's Choirs Co-operate," *Toronto Daily Star*, May 2, 1942; "Honor Two Choirs," *Toronto Daily Star*, June 13, 1942.

8 "Proud 'Little Norway' Camp Thrilled by Royal Visitors," *Evening Telegram*, January 13, 1941; "Planes at Island Airport Dip Their Wings in Salute to Royal Norwegian Couple," *Evening Telegram*, January 14, 1941; Earl Smith, "Little Norway is Inspected by Prince Olav," *Globe and Mail*, January 14, 1941.

9 "God Send Us Home in Peace Norse Prince, Airmen Pray," *Toronto Daily Star*, January 14, 1941; "Society Highlights: Toronto's Norse Residents Greet Prince and Princess," *Toronto Daily Star*, January 14, 1941.

10 "Willkie Carries Toronto Thousands Roar Acclaim," *Toronto Daily Star*, March 24, 1941; "Eyes Turn to Sky as Norse Planes Zoom," *Globe and Mail*, March 25, 1941; Saunders, *Little Norway in Pictures*, n.p.

Chapter 14

1 Looseley, "History of the Royal Norwegian Air Force in Canada," 41; "Norwegian Navy Flyers to Train at Vancouver," *Evening Telegram*, January 28, 1941; "Norse Naval Fliers to Train at Coast," *Toronto Daily Star*, January 28, 1941.

2 "Toronto Bride of 2 Months Bereaved as Two Killed in Crash of Norse Plane," *Evening Telegram*, February 22, 1941; "Crash Kills Norse Flier Who Wed Toronto Girl," *Globe and Mail*, February 22, 1941.

3 "2 Norse Fliers Killed in West," *Globe and Mail*, March 19, 1941.

4 Looseley, "History of the Royal Norwegian Air Force in Canada," 41.

5 "See Norse Fliers Swimming after Eastern Gap Crash," *Toronto Daily Star*, March 26, 1941; "Young Danish Flier Dies as Plane Dives into Lake," *Globe and Mail*, March 27, 1941; "Willkie Laments Death of Norse Airman in Crash," *Toronto Daily Star*, March 27, 1941; "Camp Stops Classes as Tribute to Flier," *Toronto Daily Star*, April 1, 1941.

6 Hansen, *Little Norway*, 48.

7 Ibid.

8 Stene, "The Distinguished Flying Cross and How It Was Won," 12, 13.

9 Timothy Allan Johnston, "Diary: No. 1 Service Flying Training School, Camp Borden," www.timothyallanjohnston.com, http://www.timothyallanjohnston.com/ (Accessed March 26, 2017). For example, the No. 1 Service Flying Training School's daily diary shows this entry for May 5, 1944: "The pupil obtaining the highest standing on course 96 was N2449 LAC Aschimm, Royal Norwegian Air Force. The second highest per centage was obtained by N2447 LAC Aagaard.... The standing achieved by LACs Aschimm and Aagaard of the Royal Norwegian Air Force in obtaining the highest percentage in this graduating course is a remark-

able achievement and testifies to their scholarly and flying ability."

10 "Cloud Combing: The Norwegians," *Evening Telegram*, February 18, 1941.

11 Hansen, *Little Norway*, 54; Looseley, "History of the Royal Norwegian Air Force in Canada," 41; "Little Norway's Graduates Busy Hunting Nazis," *Toronto Daily Star*, July 16, 1942; Norway, "The History of 'Little Norway' in Canada," 5.

12 Hansen, *Little Norway*, 65.

13 Norway, "The History of 'Little Norway' in Canada," 5, 6; "Little Norway's Graduates Busy Hunting Nazis," *Toronto Daily Star*, July 16, 1942; Saunders, *Little Norway in Pictures*, n.p.

14 Kjell Arne Reistad, Personal Interview with Author, Oslo, Norway, August 24, 2015.

15 Hansen, *Little Norway*, 62.

16 "Norwegian OC Charges Fifth Columnists Trying to Undermine Efforts," *Globe and Mail*, May 13, 1941; "Claims Norse Here 5th Column Target," *Toronto Daily Star*, May 13, 1941. Note: The rumours that members of the RNAF were engaged in subversive activities against Canada and its Allies continued to spread, and by the summer of 1941, RNAF leaders worried that the morale of the flyer trainees was being undermined and their efficiency lowered. In July 1941, RCAF Air Commodore G.E. Brookes led an inspection party that examined Little Norway's equipment and every phase of its training. Brookes spoke to the officers and men at Little Norway afterwards, praising the RNAF members and assuring them that the allegations of fifth column activities had been proven to have no basis in fact whatsoever. See "Norse Airmen Cleared of 5th Column Suspicion," *Globe and Mail*, July 16, 1941.

17 Hansen, *Little Norway*, 61.

18 "Girls Learn Norse Tongue So Can Talk to Hubbies," *Toronto Daily Star*, July 22, 1943; Eric Sanderson, "R.N.A.F. Members' Wives Learn About Life in Norway," *Globe and Mail*, August 19, 1943; "Norse Women's Auxiliary Prepares for Liberation," *Toronto Daily Star*, May 16, 1944.

19 "Norwegian Auxiliary Will Hold Carnival," *Toronto Daily Star*, October 2, 1944; Saunders, *Little Norway in Pictures*, n.p.

20 Rita Hiorth, Personal Interviews with Author, Toronto, Ontario, January 19, 2015 and March 2, 2015.

21 Sandra Howes, Personal Interview with Author, Caledon, Ontario, July 14, 2014.

22 Donald Jones, "Unravelling 'Little Norway's' Big Secrets, *Toronto Star*, May 10, 1986; Richard Ouzounian, "Scenes from a Life Lived Large," *Toronto Star*, March 14, 2015.

Chapter 15

1 Mazower, *Hitler's Empire*, 478; Riste and Nökleby, *Norway 1940–45, The Resistance Movement*, 31, 32.

2 "Reprisals Follow Raid on Islands Off Norway," *Toronto Daily Star*, March 7, 1941.

3 Eric Hansen, Email Interview with Author, January 2 and 3, 2016; Agderposten, "Honored with a Memorial," agderposten.no, http://www.agderposten.no/kjop-tilgang?aId=1.535919 (Accessed March 13, 2017).

4 "Norse Bishops Openly Attack Nazi Overlords," *Evening Telegram*, February 4, 1941; "Norse Battle 'Quislings' in Bitter Street Clashes," *Evening Telegram*, February 5, 1941.

5 Riste and Nökleby, *Norway 1940–45, The Resistance Movement*, 24.

6 Ibid., 27, 28.

7 "Norse Patriots Dare Death So People Can Get News," *Toronto Daily Star*, January 5, 1942.

8 J. Peter Hvidsten, "Per Hvidsten: An Amazing Journey, Norway to Canada," 5.

9 Inger Stene, Personal Interviews with Author, Toronto, July 17, 2014 and August 27, 2014; John Stene, "The Distinguished Flying Cross and How It Was Won," 2, 3.

Chapter 16

1 Earl Smith, "Norse Airmen Do Ski Jumps at Huntsville," *Globe and Mail*, January 13, 1941, quoting Lieutenant Rolf Berg.

2 Ibid.; Paul H. Rice, "Around Town," *Huntsville Forester*, January 9, 1941; "Norse Flyers Do Air Stunts on Their Skis," *Evening Telegram*, January 13, 1941.

3 Harold J. Briggs and Jane Briggs Van Buskirk, "The Mica Mine & Me," Fairy Lake Association, http://www.fairylakeassociation.com/history-stories-past-present/the-mica-mine-me/ (Accessed March 14, 2017).

4 "International Skiers on Sports Week-End," *Toronto Daily Star*, January 18, 1941; "Fifty Buffalo Skiers Declare First 'International Ski Week-End' at Huntsville Great Success," *Toronto Daily Star*, January 20, 1941; "Ski Train and Ski Trail Cement Anti-Axis Bonds, *Toronto Daily Star*, January 20, 1941.
5 "Snow-Clad Hills of North Delight Buffalo Skiers," *Evening Telegram*, January 20, 1941.
6 "Big Events Scheduled for Over Week-End," *Huntsville Forester*, January 30, 1941; "Carnival of Thrills and Spills as Army Ski Tourney Starts," *Toronto Daily Star*, February 1, 1941.
7 "Norwegian Fliers Clean Up in Huntsville Ski Carnival," *Toronto Daily Star*, February 3, 1941; "Ski Troops from Many Lands Compete at Huntsville Meet," *Toronto Daily Star*, February 3, 1941.
8 "Norwegian Fliers Clean Up in Huntsville Ski Carnival," *Toronto Daily Star*, February 3, 1941.
9 Ibid.; "Norwegian Airmen Feature First Military Ski Meet," *Evening Telegram*, February 3, 1941; "Ski Troopers Feature of Huntsville Tourney," *Globe and Mail*, February 3, 1941; "First Military Ski Meet is a Huge Success," *Huntsville Forester*, February 6, 1941.
10 "Soccer Team to Chicago," *Globe and Mail*, June 12, 1942; Looseley, "History of the Royal Norwegian Air Force in Canada," 22.
11 "Norwegian Keeps Date and Nearly Misses Baton!" *Toronto Daily Star*, June 5, 1942.
12 "Largest Parade in City Since First Great War Watched by 100,000," *Evening Telegram*, February 22, 1941; Earl Smith, "100,000 Line Streets to See Five-Mile Parade of Troops and Civilians Spur Savings Drive," *Globe and Mail*, February 24, 1941.
13 Earl Smith, "Campaign Success Sure, Celebrate on Saturday," *Globe and Mail*, March 14, 1941.
14 "Over the Teacups," *Toronto Daily Star*, December 18, 1940.
15 "Little Norway Does Big Share in First Year," *Toronto Daily Star*, November 11, 1941.
16 "Allies May Invade Norway 'Any Time Now,' Prince Says," *Toronto Daily Star*, January 20, 1942; "Spirit of 'Little Norway,'" Card, Toronto, n.d., Large Hol. Box 24, Gravenhurst Archives, Gravenhurst, Ontario; Looseley, "History of the Royal Norwegian Air Force in Canada," 23, 24.
17 Looseley, "History of the Royal Norwegian Air Force in Canada," 26.
18 Carl Sandburg, "Carl Sandburg Says," *Wings for Norway*, 1942, 25, Karen Greavette Collection.
19 "Southern Norsemen Send Plane as Gift," *Toronto Daily Star*, March 28, 1941.
20 "Havergal Student Guest of Royalty," *Toronto Daily Star*, January 20, 1942.
21 "Norway Raids Boost Morale, Prince States," *Globe and Mail*, January 21, 1942.
22 "Ten Norwegian War Heroes Fly from Toronto to White House," *Toronto Daily Star*, September 25, 1941; "Ten Norwegians Leave Toronto for Good-Will Tour of U.S.," *Toronto Daily Star*, September 25, 1941; "Sabotage Cheered by Norse Princess," *Toronto Daily Star*, September 27, 1941.
23 "Stamp Corner," *Toronto Daily Star*, January 23, 1943; "Stamp to Honor 'Little Norway,' " *Globe and Mail*, February 13, 1946; C.R. McGuire, "The 'Little Norway' Story," History, Ottawa, 1980, 9–11, Large Hol. Box 47B, Gravenhurst Archives, Gravenhurst, Ontario.

Chapter 17

1 Looseley, "History of the Royal Norwegian Air Force in Canada," 8.
2 W.J.E. Johnston to E.L. Cousins, Letter, April 1, 1941, RG 3/3, Box 33, PortsToronto Archives, Toronto, Ontario.
3 E.L Cousins to Mayor Conboy, Letter, April 17, 1941, RG 3/3, Box 33, PortsToronto Archives, Toronto, Ontario; E. L. Cousins to G.K. Sheils, Letter, January 28, 1942, RG 3/3, Box 33, PortsToronto Archives, Toronto, Ontario.
4 Toronto Harbour Commissioners, Stenographic Report of Presentation Made by a Deputation from the Royal Norwegian Air Force, May 21, 1941, RG 3/3, Box 33, PortsToronto Archives, Toronto, Ontario; Royal Norwegian Air Force to Toronto Harbour Commissioners, Submission, May 28, 1941, RG 3/3, Box 33, PortsToronto Archives,Toronto, Ontario.
5 F.R. Scandrett to W. Zimmerman, Letter, June 30, 1941, RG 3/3, Box 33, PortsToronto Archives, Toronto, Ontario; William Zimmerman to Toronto Harbour Commissioners, Letter, July 12, 1941, RG 3/3, Box 33, PortsToronto Archives, Toronto, Ontario.
6 "Oil Patch on Lake May Hide Plane," *Toronto Daily Star*, April 19, 1941; "Rain Hampers Norse Search," *Globe and Mail*, April 21, 1941; "Norse

Flier, Plane Sought by Nearly 30 War Craft," *Toronto Daily Star*, April 24, 1941; "Must Find Norse Flier in 3 Days or 'He's Dead'," *Toronto Daily Star*, April 25, 1941; "Hope Ebbs Pilot XXX Will Be Found Alive," *Toronto Daily Star*, April 26, 1941.

7 Harald Olafson, "Little Norway in Muskoka," *The Light of Other Days*, 2nd ed., Gravenhurst Historical Committee (Cobourg, Ontario: Gravenhurst Historical Committee, 1972), 126; North, "Beyond the Seas," 317, 318.

8 "Funeral is Held for Two Fliers," *Globe and Mail*, May 2, 1941.

9 "Little Norway Locates Camp at Emsdale," *Huntsville Forester*, July 10, 1941.

10 Stene, "The Distinguished Flying Cross and How It Was Won," 13.

11 Rolf Kolling, Personal Interview with Author, Oslo, Norway, August 25, 2015.

12 "Hills, Bush Recall Native Land to Norse Training at Emsdale," *Toronto Daily Star*, October 11, 1941; Stene, "The Distinguished Flying Cross and How It Was Won," 13.

13 Jensen, *The Adrenalin Years*, 72.

14 "Countryside News," *Huntsville Forester*, October 2, 1941.

15 "Around Town," *Huntsville Forester*, July 31, 1941.

Chapter 18

1 "Summer Hotel Deep in Bush Serves Norse Fliers Only," *Toronto Daily Star*, July 12, 1941; "Summer Hotel for Norwegians," *Gravenhurst Banner*, August 21, 1941; Mary-Lyn Tebby, "Vesle Skaugum: Oxbow Lake's Norwegian Past," *Muskokan*, August 5, 1999; Instrument 6328, Grant, The Crown to Frank Peelar, Registered September 11, 1912, Muskoka Land Registry Office #35, Bracebridge, Ontario.

2 "Hills, Bush Recall Native Land to Norse Training at Emsdale," *Toronto Daily Star*, October 11, 1941; "Norse in Ontario Honor Prince Olav and Wife," *Globe and Mail*, January 20, 1942; Penny Gumbert, "Muskoka Hospitable to Norwegian Servicemen," *Muskoka Magazine*, October/November 2005, 56; Looseley, "History of the Royal Norwegian Air Force in Canada," 4, 10, 11, 26.

3 "Dynamite Artist in Norway Builds Norse Air Base Here," *Toronto Daily Star*, March 4, 1942.

4 Scott McClellan, "Recalling the Spirit of Little Norway, Part II: Toronto," *Muskoka Sun*, June 13, 1985.

5 Brad Hammond, "In Their Own Words: The Vikings in Muskoka," *Vintage Muskoka*, May 2002, 17, File 05-45, Gravenhurst Archives, Gravenhurst, Ontario.

6 Eva-Lis Wuorio, "New Lodge in Ontario Woods Prepares Norse for War," *Evening Telegram*, January 31, 1942.

7 Per Falkenhall Jr., Telephone Interview with Author, March 27, 2015.

8 "Norwegian Royalty Skiing at Huntsville," *Toronto Daily Star*, January 19, 1942; Dennis Stone, "R.N.A.F. Bases in Muskoka During World War II," *Muskoka Sun*, August 23, 1973.

9 "Allies May Invade Norway 'Any Time Now', Prince Says," *Toronto Daily Star*, January 20, 1942.

10 Eva-Lis Wuorio, "New Lodge in Ontario Woods Prepares Norse for War," *Evening Telegram*, January 31, 1942.

11 "Norse Fliers Tested in 20-Mile Ski Trek," *Toronto Daily Star*, March 9, 1942.

12 "Commanding Officer Leads His Men Home," *Toronto Daily Star*, March 9, 1942; "Ole Reistad Tops Field in Ski Test," *Globe and Mail*, March 9, 1942; "Lieut.-Col. Ole Reistad Winner of Long Ski Race," *Huntsville Forester*, March 12, 1942.

13 Saunders, *Little Norway in Pictures*, n.p.; "Norwegians Complete Long March—Camp Over Night at Lansing," Article, Anonby Family Collection.

14 Hansen, *Little Norway*, 86–88.

15 "Norwegian Skier Wins U.S. Title," *Toronto Daily Star*, February 9, 1942; "Ola Adds to Ski Laurels," *Globe and Mail*, February 9, 1942; "Here Comes Ola in Action Shots Showing Ski-Jumping Made Easy," *Toronto Daily Star*, February 14, 1942.

16 Jensen, *The Adrenalin Years*, 67.

17 Paul Morton, "Norse Fliers Sure Now That 5-Year-Old Prince is True Son of Vikings," *Toronto Daily Star*, September 23, 1942; "Prince Harald of Norway Wins Hearts of Fighters Training Near Huntsville," *Evening Telegram*, September 23, 1942; "Youthful Prince Harald of Norway Visits His Airmen and Proves Real 'Prince Charming' to the Entire Camp," *Evening Telegram*, September 24, 1942.

Chapter 19

1 "Think Dazzling Sunlight Caused Norse Fliers' Crash," *Toronto Daily Star*, June 21, 1941;. "Coroner Fears Airplane May Fall in City Street if Flying Not Controlled," *Evening Telegram*, July 12, 1941.
2 "Think Dazzling Sunlight Caused Norse Fliers' Crash," *Toronto Daily Star*, June 21, 1941.
3 "Stress Danger Novice Pilots are to Citizens," *Globe and Mail*, July 12, 1941; "Would Not Happen Again in 100 Years—Norseman," *Toronto Daily Star*, July 12, 1941.
4 "Would Not Happen Again in 100 Years—Norseman," *Toronto Daily Star*, July 12, 1941.
5 Ibid.; "Stress Danger Novice Pilots are to Citizens," *Globe and Mail*, July 12, 1941.
6 "Seasonable Word of Warning from the Chief Coroner," *Evening Telegram*, July 14, 1941.
7 "4 Airmen Killed Here When 2 Bombers Crash," *Globe and Mail*, July 7, 1941; "Norsemen Hurtle to Death Amid Fiery Plane Wreckage," *Toronto Daily Star*, July 5, 1941.
8 "Thousands on Beaches Watch Norwegian Flier Crash to Death in Lake," *Globe and Mail*, July 28, 1941; "Norse Flier Falls in Lake Before Eyes of Thousands," *Toronto Daily Star*, July 28, 1941.
9 Reed-Olsen, *Two Eggs on My Plate*, 92.
10 "Little Norway Does Big Share in First Year," *Toronto Daily Star*, November 11, 1941.
11 "Norse Fliers Win R.A.F. Praise for Daredevil Daylight Raids," *Toronto Daily Star*, December 23, 1941.
12 Ibid.; "'Grads' of Little Norway Earn Praise Overseas," *Globe and Mail*, December 24, 1941; "Anonymous Norsemen Play Great Part in War, *Globe and Mail*, March 24, 1942; "Little Norway's Graduates Busy Hunting Nazis," *Toronto Daily Star*, July 16, 1942.
13 "Norse Fliers Win R.A.F. Praise for Daredevil Daylight Raids," *Toronto Daily Star*, December 23, 1941; Cato Guhnfeldt, *Spitfire Saga*, vol. 1, 116, 124.
14 Kai Pedersen, "Little Norway May She Always Be Free," *Forester*, August 19, 1981; Looseley, "History of the Royal Norwegian Air Force in Canada," 50.
15 "Norse Plane in Formation Spins to Earth, Pilot Dies," *Toronto Daily Star*, January 21, 1942; "Norwegian Airman Dies in Crash Near Oakville," *Globe and Mail*, January 21, 1942; "Norse Airman Dies in Crash in Mock Fight," *Evening Telegram*, January 21, 1942; "Lack of 300 Feet Spelt Death for Norse in 400 m.p.h. Dive," *Toronto Daily Star*, January 22, 1942.
16 "Saw 'Chute Drop from Sky 'Must Have Gone Into Lake,' " *Toronto Daily Star*, January 26, 1942; "2-Day Hunt Fails to find Norse Pilot and Plane," *Globe and Mail*, January 26, 1942; "Plane Search Still Fruitless," *Globe and Mail*, January 27, 1942; "Quit Search for Plane and Pilot, Lieut. T. Nalki," *Globe and Mail*, January 28, 1942; "Norwegian Hero," *Toronto Daily Star*, January 28, 1942.
17 "Norse Flier, 23, Killed in Mid-Air Collision," *Globe and Mail*, July 23, 1942; "Planes Collide in Mid-Air Norwegian Flier is Killed," *Toronto Daily Star*, July 23, 1942; "Planes Crash in Mid-Air One Killed, Two Unhurt," *Evening Telegram*, July 23, 1942.
18 "Deaths," *Toronto Daily Star*, October 3, 1942; Hansen, *Little Norway*, 104.
19 "Over the Teacups," *Toronto Daily Star*, February 13, 1942; "Land Plane in Field 'Could Have Crashed,' " *Toronto Daily Star*, February 12, 1942.

Chapter 20

1 Muskoka Airport, "History," Muskoka Airport, http://www.muskokaairport.com/history/ (Accessed March 16, 2017).
2 Reay Women's Institute, "Current Events (Dominion Airport)," "Tweedsmuir Village History," History, 37, Gravenhurst Archives, Gravenhurst, Ontario.
3 Muskoka Airport, "History," Muskoka Airport, http://www.muskokaairport.com/history/ (Accessed March 16, 2017).
4 Albert Jardine, "Muskoka Airport, Part 3," *Gravenhurst News*, September 29, 1977.
5 Per Conradi Hansen, "The Royal Norwegian Air Force in Canada, 1940–1945," *Canadian Defence Quarterly*, vol. 17, no. 2, Autumn 1987, 63.
6 Hansen, *Little Norway*, 54, 66.
7 Looseley, "History of the Royal Norwegian Air Force in Canada," 11; Instrument 4612, Grant, Thomas Short to Edvard Omholt-Jensen, Registered February 6, 1942, Muskoka Land Registry Office #35, Bracebridge, Ontario.

8 "Norwegian Trainees Moving to Muskoka," *Toronto Daily Star*, February 21, 1942; "Norwegians Move to Muskoka," *Huntsville Forester*, March 5, 1942; "Norwegian Air Force Camp to Be Moved to Muskoka," Article, Anonby Family Collection.

9 "Dynamite Artist in Norway Builds Norse Air Base Here," *Toronto Daily Star*, March 4, 1942.

10 Danish WW2 Pilots, "Bourbon de Parma, Prince Jacques de (1922–1964)," Danish WW2 Pilots, http://www.danishww2pilots.dk/profiles.php?id=248 (Accessed March 16, 2017).

11 Brad Hammond, "Room for the Sky and All Its Stars: Muskoka's Thor Heyerdahl," *Vintage Muskoka*, approx. 2001, 26–31; Axel Andersson, *Hero for the Atomic Age: Thor Heyerdahl and the Kon-Tiki Expedition* (Oxford, United Kingdom: Peter Lang Ltd., 2010), 24–28.

12 Patricia Pedersen, Telephone Interviews with Author, March 3 and 31, 2015.

Chapter 21

1 Riste and Nökleby, *Norway 1940–45*, 40.

2 Ibid., 41–43; "Herd 500 Norse Teachers in Ship Crowded by 250," *Toronto Daily Star*, April 27, 1942; "8 Months of Living Death Can't Break Norse Exiles," *Toronto Daily Star*, May 12, 1943.

3 "Only 18 Norse Clergy Quisling 1,500 Quit Jobs," *Toronto Daily Star*, March 4, 1942; Riste and Nökleby, *Norway 1940–45*, 44; "Quisling Rules Churches Only 30 of 1,000 Are Open," *Toronto Daily Star*, April 14, 1942.

4 United States Holocaust Memorial Museum, "Norway," www.ushmm.org, https://www.ushmm.org/wlc/en/article.php?ModuleId=10005460 (Accessed March 16, 2017); BBC News, "Norway Apologizes for Deporting Jews During Holocaust," BBC News Services, http://www.bbc.com/news/world-europe-16761558 (Accessed March 16, 2017).

5 Riste and Nökleby, *Norway 1940–45*, 46, 47.

6 Geoff Ward, "The Tempsford Taxis," Norway: War, Resistance, Peace, http://www.wwiinorge.com/our-stories/the-tempsford-taxis/ (Accessed February 7, 2017).

7 Ibid.

8 Riste and Nökleby, *Norway 1940–45*, 51, 52.

9 Robert McG. Thomas Jr., "Max Manus, 81, Dies: Fought Nazis in Occupied Norway," *New York Times*, September 22, 1996; Bob Pearson, "Max Manus—An Extraordinary Man," Norway During World War 2, http://www.nuav.net/maxmanus.html (Accessed March 16, 2017). On November 26, 1942, the *Donau* had transported Jews from Norway to Germany.

10 Sønsteby, *Report from #24*, 11–100.

11 Reed-Olsen, *Two Eggs on My Plate*, 92, 96, 102–106.

Chapter 22

1 "Little Norway's Graduates Busy Hunting Nazis," *Toronto Daily Star*, July 16, 1942.

2 Rolf Hauge, Personal Interview with Author, Collingwood, Ontario, May 6, 2015.

3 Norway, "The History of 'Little Norway' in Canada," 5, 6.

4 Rolf Kolling, Personal Interview with Author, Oslo, Norway, August 25, 2015; Telephone Interview with Author, January 16, 2015.

5 Knut Fossum, Personal Interview with Author, Oslo, Norway, August 25, 2015.

6 "Little Norway's Graduates Busy Hunting Nazis," *Toronto Daily Star*, July 16, 1942; Guhnfeldt, *Spitfire Saga*, vol. 1, 287.

7 "Fled Norway, Trained Here, Missing After Air Fight," *Toronto Daily Star*, January 9, 1942; "Norse Airman Listed Missing," *Globe and Mail*, January 10, 1942.

8 Jon Guttman, "Norway's Ace of Aces," *Aviation History*, July 1999, 44.

9 Larsen and Thorsager, *Viking Spitfire*, 110, 111.

10 "Flyers from Little Norway Down 14 Nazis at Dieppe," *Evening Telegram*, August 28, 1942; Guhnfeldt, *Spitfire Saga*, vol. 2, 239.

11 "Flyers from Little Norway Down 14 Nazis at Dieppe," *Evening Telegram*, August 28, 1942; "3 Norwegian Fliers are Awarded D.F.C.," *Toronto Daily Star*, September 5, 1942.

12 "Little Norway Men Fought at Dieppe," *Globe and Mail*, August 27, 1942.

13 "3 Norwegian Fliers are Awarded D.F.C.," *Toronto Daily Star*, September 5, 1942; "Norse Airmen Awarded D.F.C.," *Globe and Mail*, September 7, 1942.

14 "Ferry Fliers Outgrow Port at Montreal," *Globe and Mail*, July 27, 1942.

15 "Norway Fliers Trained Here Now in Hot Pursuit of Nazis," *Toronto Daily Star*, August 7, 1942; "Nazis Feel Vengeance of Norwegian Airmen Who Graduated Here," *Evening Telegram*, August 7, 1942; "Don't Want

to Return as Instructors, Fighting Norse Declare Otherwise," *Globe and Mail*, August 7, 1942.

16 Stene, "The Distinguished Flying Cross and How It Was Won," 14–16; Spitfirepilots, "Per Waaler—Gloster Gladiator Pilot," Spitfirepilots.com, http://www.spitfirepilots.com/per-waaler-gloster-gladiator-pilot/ (Accessed March 16, 2017).

Chapter 23

1 "Dynamite Artist in Norway Builds Norse Air Base Here," *Toronto Daily Star*, March 4, 1942.

2 John Clement, Email Interview with Author, June 3, 2014.

3 "New Norwegian Air Camp Is Opened at Gravenhurst," *Toronto Daily Star*, May 4, 1942.

4 "Norse Prince Dedicates New R.N.A.F. School," *Globe and Mail*, May 5, 1942.

5 Ken W. MacTaggart, "Latest 'Little Norway' Gives Men Home Flavor," *Globe and Mail*, March 10, 1942; "Royal Norwegian Couple Given Rousing Welcome By Gravenhurst Throng," *Evening Telegram*, May 5, 1942.

6 Reay Women's Institute, "Current Events (Dominion Airport)," 38; Catherine Groth Sparrow, "The Spirit of Little Norway—Part II, *Muskoka Sun*, June 25, 1981; Looseley, "History of the Royal Norwegian Air Force in Canada," 11.

7 Sparrow, "The Spirit of Little Norway—Part II, *Muskoka Sun*, June 25, 1981.

8 Ken W. MacTaggart, "Latest 'Little Norway' Gives Men Home Flavor," *Globe and Mail*, March 10, 1942.

9 "Royal Norwegian Couple Given Rousing Welcome By Gravenhurst Throng," *Evening Telegram*, May 5, 1942; "Norwegian Royalty Visit Gravenhurst," *Gravenhurst Banner*, May 7, 1942.

10 Examples of newspapers referring to the camp as "Little Bergen": "New Norwegian Air Camp Is Opened at Gravenhurst," *Toronto Daily Star*, May 4, 1942; "Royal Visitors Attend Official Opening of New Norwegian Camp in Muskoka," *Evening Telegram*, May 4, 1942; "Little Bergen Dedicated by Norwegians," *Globe and Mail*, May 5, 1942.

11 Hansen, *Little Norway*, 79; "Crown Prince Finds Laughter in Camp Name," *Evening Telegram*, May 5, 1942.

12 Sparrow, "The Spirit of Little Norway—Part II, *Muskoka Sun*, June 25, 1981.

13 Obituary, Per Kaare Nilsen, Gravenhurst Archives, Gravenhurst, Ontario. Later, Nilsen volunteered for overseas duty and served in the United Kingdom, France, Belgium, and Holland. For a period after the war, Nilsen was employed as a chef at Gravenhurst's Gilmour Hotel, where crowds came on Sunday nights to enjoy his famous smorgasbord dinners.

14 Looseley, "History of the Royal Norwegian Air Force in Canada," 32, 33.

15 Shirley Barlow, Telephone Interviews with Author, May 21, 2014 and November 7, 2014.

16 "Norse Prince Dedicates New R.N.A.F. School," *Globe and Mail*, May 5, 1942.

17 Thomas Rennie to Ole Reistad, Letter, March 20, 1943, RG 3/3, Box 33, PortsToronto Archives, Toronto, Ontario.

18 Ole Reistad to Thomas Rennie, Letter, March 30, 1943, RG 3/3, Box 33, PortsToronto Archives, Toronto, Ontario.

19 W.J.E. Johnston to Alex McLeod, Letter, March 10, 1943, RG 3/3, Box 33, PortsToronto Archives, Toronto, Ontario; Alex McLeod to W.J.E. Johnston, Letter, March 10, 1943, RG 3/3, Box 33, PortsToronto Archives, Toronto, Ontario; W.J.E. Johnston to Acting General Manager, Letter, March 11, 1943, RG 3/3, Box 33, PortsToronto Archives, Toronto, Ontario; W.J.E. Johnston to Alex McLeod, Letter, March 12, 1943, RG 3/3, Box 33, PortsToronto Archives, Toronto, Ontario.

20 E.L. Cousins to Alex McLeod, Letter, March 31, 1943, RG 3/3, Box 33, PortsToronto Archives, Toronto, Ontario.

21 "Little Norway Air School Turned Over to R.C.A.F.," *Toronto Daily Star*, April 6, 1943; "Toronto Little Norway Taken Over by R.C.A.F.," *Globe and Mail*, April 6, 1943.

22 Ken W. MacTaggart, "Bright Record Written by Airmen," *Globe and Mail*, March 9, 1943.

23 "Little Norway Doors to Open," *Globe and Mail*, October 7, 1944.

Chapter 24

1 "Rough, Tough Foes of Nazis are Norse Commando Dogs," *Toronto Daily Star*, February 11, 1943.

2 Given the era, it was not unusual that the RNAF raised orphaned bear

cubs. For instance, until recent times, Canadian and American highway gas stations and gift shops sometimes kept black bears in cages as attractions to bring in customers.

3 Betty Honnor, Personal Interview with Author, Streetsville, Ontario, June 11, 2014; Larry Baadsvik, Personal Interview with Author, Gravenhurst, Ontario, October 17, 2016.

4 "Sonja Flies to Muskoka to Visit Norse Airmen," *Globe and Mail*, March 23, 1944; "Sonja Gets Rubber Boots to Tour Little Norway," *Toronto Daily Star*, March 23, 1944.

5 Randi Hvatum, Telephone Interview with Author, May 16, 2014; Harald Olafson, "Little Norway in Muskoka," *Light of Other Days*, 125.

6 Kjell Arne Reistad, Personal Interview with Author, Oslo, Norway, August 24, 2015; "'We're Coming Back Again,' is Refrain as Norwegians Break Camp, Say Farewell," *Toronto Daily Star*, February 17, 1945; "Fanny and Her Furry Family Will Shock Norwegian Flier," *Globe and Mail*, August 23, 1945.

7 Douglas Johnston, Email Interview with Author, January 6, 2016.

8 Douglas Johnston, Personal Interview with Author, Burlington, Ontario, February 8, 2016.

9 Hansen, *Little Norway*, 100, 112, 113.

10 Nancy Lynch, Telephone Interview with Author, June 19, 2014.

11 Yvonne Jones, Personal Interview with Author, Gravenhurst, Ontario, July 2014.

12 "Restaurant Closes on Wednesdays," *Gravenhurst Banner*, May 13, 1943.

13 Ron Stoddart, Telephone Interviews with Author, May 23, 2014 and February 10, 2017.

14 Hansen, *Little Norway*, 113, 116.

15 "Blond Babies Soon May See Land of Daddy," *Toronto Daily Star*, January 11, 1945.

16 Barbara Chramer, Personal Interview with Author, Caledon, Ontario, June 18, 2015.

Chapter 25

1 Cecil Porter, *The Gilded Cage: Gravenhurst German Prisoner-of-War Camp 20, 1940–1946*, 2nd ed. (Gravenhurs, Ontario: Gravenhurst Book Committee, 2003), 43.

2 Paul McLaughlin, "The War Years," *Cottage Life*, May/June 1991, 98.

3 Kaye Kruusmagi, Personal Interview with Author, Mississauga, Ontario, January 14, 2016; "Holiday-Makers Pass Prison Camp on Luxurious Vessel with Band," *Evening Telegram*, August 17, 1940.

4 "Norwegian Air Force Camp to be Moved to Muskoka," Article, Anonby Family Collection.

5 Scott McClellan, "Recalling the Spirit of Little Norway, Part III: Muskoka," *Muskoka Sun*, June 20, 1985.

6 Hansen, *Little Norway*, 113.

7 Kaye Kruusmagi, Personal Interview with Author, Mississauga, Ontario, January 14, 2016; "Norwegian Air Force Trainees Join Hunt for Escaped Nazi," *Toronto Daily Star*, December 10, 1942; "Missing Prisoner Found in Camp," *Gravenhurst Banner*, April 8, 1943; Porter, *The Gilded Cage*, 141, 142.

8 Porter, *The Gilded Cage*, 148, 149.

9 L. Gold, "Know Your Hospital," *The Sanitarium Sun*, February 1954, 13. At least two Norwegian flyers died of tuberculosis at Muskoka Hospital.

10 Christine Cator Everest, Personal Interview with Author, Gravenhurst, Ontario, May 15, 2014.

11 Earle Laycock, Telephone Interview with Author, January 4, 2016.

12 "'For Norway,' Is War Cry of Viking Air Force Girls," *Toronto Daily Star*, November 16, 1943; "First Wren from Norway Longs to Be Back on Skis," *Toronto Daily Star*, March 15, 1944; Looseley, "History of the Royal Norwegian Air Force in Canada," 28.

13 Sandra Howes, Personal Interview with Author, Caledon, Ontario, July 14, 2014.

14 "'For Norway,' Is War Cry of Viking Air Force Girls," *Toronto Daily Star*, November 16, 1943.

15 "New Garb Styled at Little Norway," *Globe and Mail*, November 17, 1943; Looseley, "History of the Royal Norwegian Air Force in Canada," 28.

16 "'Victory Train' Visits Gravenhurst," *Gravenhurst Banner*, February 12, 1942; "Army Train Visits Bracebridge," *Bracebridge Gazette*, February 12, 1942.

17 "'Victory Parade' to Entertain Little Norway," *Gravenhurst Banner*, August 5, 1943; "Coca-Cola Co. Gives Stage Show for the R.N.A.F.," *Gravenhurst Banner*, August 12, 1943.

18 "Salute to the Guard and Grand Parade Monday, May 24th," *Gravenhurst Banner*, May 20, 1943; "Veterans Guard Honored Monday," *Gravenhurst*

Banner, May 27, 1943; "Gravenhurst Celebrates Victoria Day," *Huntsville Forester*, May 27, 1943; Cecil Porter, "On the Homefront," *Muskoka Today*, May 7, 1995.

19 Hansen, *Little Norway*, 97; "Deaths: Selmer," *Toronto Daily Star*, January 15, 1943.

20 "Norway Flier in Crash Fog Hid Rest of Flight," *Toronto Daily Star*, January 25, 1943; "Norwegian Airman is Killed at Norval," *Globe and Mail*, January 25, 1943.

21 "Roars at 300 M.P.H. to Earth and Death," *Toronto Daily Star*, February 23, 1943; "Norway Flier Dies in Crash," *Globe and Mail*, February 23, 1943; "Crash Victim's Wife, Son Free," *Globe and Mail*, February 24, 1943.

22 Kai Pedersen, "Little Norway May She Always Be Free," *Forester*, August 19, 1981; "Aeroplane Crash at Germania," *Bracebridge Gazette*, August 31, 1944; Hansen, *Little Norway*, 117.

23 Reay Women's Institute, "Current Events (Dominion Airport)," 38; "Aeroplane Crash at Germania," *Bracebridge Gazette*, August 31, 1944; "Aeroplane Crash at Reay," *Bracebridge Gazette*, October 19, 1944; Hansen, *Little Norway*, 117. Svend Clement to Editor of *Gravenhurst Banner*, Letter, n.d.

Chapter 26

1 "The Drama of Heavy Water," (Brochure, Norway, n.d., Bacon Family Collection); Riste and Nökleby, *Norway 1940–1945*, 59–61.

2 Reed-Olsen, *Two Eggs on My Plate*, 134–203, 211, 212, 219–221.

3 "Nazis Sending Families Home Norse Situation 'Near Explosion,'" *Toronto Daily Star*, October 17, 1944; Sønsteby, *Report from #24*, 107–111; Riste and Nökleby, *Norway 1940–1945*, 73–75.

4 Sønsteby, *Report from #24*, 130–134, 153, 161.

5 Riste and Nökleby, *Norway 1940–1945*, 83.

Chapter 27

1 Stene, "The Distinguished Flying Cross and How It Was Won," 16–20.

2 Scott McClellan, "Recalling the Spirit of Little Norway, Part II: Toronto," *Muskoka Sun*, June 13, 1985.

3 Stene, "The Distinguished Flying Cross and How It Was Won," 20, 21.

4 Canada, "D-Day and the Battle of Normandy," Veterans Affairs Canada, http://www.veterans.gc.ca/eng/remembrance/history/second-world-war/d-day (Accessed March 18, 2017).

5 Wilhelm Mohr, "The Contribution of the Norwegian Air Forces," *Britain and Norway in the Second World War*, ed. Patrick Salmon (London, England: H.M. Stationery Office, 1995), 91, 92; Wilhelm Mohr, "The Ground Crews' Participation in the War on the Continent 1944/45," *Wilhelm Mohr on World War II*, 81–83.

6 Cato Guhnfeldt, *Spitfire Saga*, vol. 5, 287.

7 Canada, "D-Day and the Battle of Normandy," Veterans Affairs Canada, http://www.veterans.gc.ca/eng/remembrance/history/second-world-war/d-day (Accessed March 18, 2017).

8 Wilhelm Mohr, "The Ground Crews' Participation in the War on the Continent 1944/45," *Wilhelm Mohr on World War II*, 83, 84.

9 Jenkins, *Churchill: A Biography*, 745; Canada, "D-Day and the Battle of Normandy," Veterans Affairs Canada, http://www.veterans.gc.ca/eng/remembrance/history/second-world-war/d-day (Accessed March 18, 2017).

10 Gudbrandsdølen Dagningen, "Rolf Kolling (92) Met with President Obama," GD, http://www.gd.no/nyheter/rolf-kolling-92-motte-president-obama/s/1-934610-7404817 (Accessed March 18, 2017).

11 Cato Guhnfeldt, *Spitfire Saga*, vol. 5, 337.

12 Rolf Kolling, Personal Interview with Author, Oslo, Norway, August 25, 2015.

13 Per Falkenhall Jr., Telephone Interview with Author, March 27, 2015.

14 Hansen, *Little Norway*, 145, 149.

15 "Catalina-D-FP 183 *Vingtor II*," Article, Little Norway Memorial, Muskoka Airport, Gravenhurst; Rolf Hauge, Personal Interview with Author, Collingwood, Ontario, May 6, 2015.

16 Rolf Hauge, Personal Interview with Author, Collingwood, Ontario, May 6, 2015; Dave Brown, "The Faithful Cat That Saved a Stranded Crew," *Ottawa Citizen*, October 8, 1999; "Under the Midnight Sun...," *Vintage Muskoka*, May 2002, 8, 9, File 05-45, Gravenhurst Archives, Gravenhurst, Ontario.

17 Larsen and Thorsager, *Viking Spitfire*, 165–170.

18 "Norsemen Greet with Joy Soviet Liberating Forces," *Toronto Daily Star*, October 27, 1944; "Norsemen Invade Homeland to Help Russians Oust Foe," *Toronto Daily Star*, November 14, 1944; The Royal House of Norway,

"Norway Thanks You," RoyalCourt.no, http://www.royalcourt.no/nyhet.html?tid=123132&sek=27262 (Accessed March 18, 2017).

Chapter 28

1 Wilhelm Mohr, "The Contributions of the Norwegian Air Force," *Wilhelm Mohr on World War II*, 41.

2 "'We're Coming Back Again' is Refrain as Norwegians Break Camp, Say Farewell," *Toronto Daily Star*, February 17, 1945.

3 Rita Hiorth, Personal Interview with Author, Toronto, Ontario, March 2, 2015.

4 Trent Frayne, "Grateful Forever to Canadians, Says Commander," *Globe and Mail*, February 17, 1945; "Ever Grateful Declare Norse Leaving Canada," *Toronto Daily Star*, February 17, 1945; "Norwegians Say Farewell," *Globe and Mail*, February 19, 1945.

5 "Little Norway Bids Farewell to Canada," *Huntsville Forester*, February 22, 1945.

6 "Farewell Ceremonies Held At Little Norway," *Gravenhurst Banner*, February 22, 1945; Cecil Porter, "On the Home Front," *Muskoka Today*, May 7, 1995.

7 "Farewell Ceremonies Held At Little Norway," *Gravenhurst Banner*, February 22, 1945.

8 "Officers' Mess Partly Destroyed," *Gravenhurst Banner*, February 22, 1945; "Serious Fire at Camp Little Norway," *Bracebridge Gazette*, February 23, 1945; Looseley, "History of the Royal Norwegian Air Force in Canada," 11.

9 Looseley, "History of the Royal Norwegian Air Force in Canada," 41, 43.

Chapter 29

1 Nils Christensen, Letter, n.d., Little Norway Memorial, Muskoka Airport, Gravenhurst, Ontario.

2 Riste and Nökleby, *Norway 1940–45*, 82–87.

3 Gunhild Ramm Reistad, Personal Interview with Author, Oslo, Norway, August 24, 2015.

4 Riste and Nökleby, *Norway 1940–45*, 88, 89.

5 Cecil Porter, "On the Homefront," *Muskoka Today*, May 7, 1995.

6 Ibid.; "Gravenhurst Celebrates Defeat of Nazi Germany," *Gravenhurst Banner*, May 10, 1945.

7 "Monday's Celebration," *Bracebridge Gazette*, May 10, 1945.

8 Ibid.

9 "Gravenhurst Celebrates Defeat of Nazi Germany," *Gravenhurst Banner*, May 10, 1945.

10 Ibid.; Cecil Porter, "On the Homefront," *Muskoka Today*, May 7, 1995.

11 "Huntsville Hysterically Happy," *Huntsville Forester*, August 16, 1945.

12 "County Town Celebrates Fall of Japan," *Bracebridge Gazette*, August 16, 1945.

13 "Two Camps," *The Light of Other Days*, 73; "Defeat of Japan Again Brings Peace to War-Torn World," *Gravenhurst Banner*, August 16, 1945.

Chapter 30

1 Rolf Kolling, Telephone Interview with Author, January 18, 2015.

2 Per Falkenhall Jr., Telephone Interview with Author, March 27, 2015.

3 Looseley, "History of the Royal Norwegian Air Force in Canada," 92.

4 Sønsteby, *Report from #24*, 190.

5 Looseley, "History of the Royal Norwegian Air Force in Canada," 22.

6 "Norwegian Airmen Ready for Invasion of Their Canadian Wives and Children," *Globe and Mail*, August 24, 1945.

7 "Canadian Brides Off for Norway," *Toronto Daily Star*, September 24, 1945.

8 "Studies Norse Year She's on Way 'Home'," *Toronto Daily Star*, October 27, 1945.

9 Patricia Palmer, "Norway Has a 'Little Canada' Now, Many of the Girls from Toronto," *Globe and Mail*, December 17, 1946.

10 Bill Whiteside and Susan Whiteside Davis, Personal Interview with Author, Toronto, Ontario, July 22, 2015.

11 Barbara Chramer, Personal Interview with Author, Caledon, Ontario, June 18, 2015.

12 Laura Anonby and Lauretta Anonby, Email Interviews with Author, July 15, 2015 and July 30, 2015.

13 "Canadian Brides Have Housing Trouble in Oslo," *Globe and Mail*, June 11, 1946.

14 "Canadian Brides Happy in Norway, Says War Flier," *Globe and Mail*,

June 21, 1946.

15 Looseley, "History of the Royal Norwegian Air Force in Canada," 11.

16 "W.H. Doherty Purchases Airport," *Gravenhurst Banner*, August 23, 1945; "Camp Little Norway Is Sold; Becomes Civilian Airfield," *Globe and Mail*, August 23, 1945; Instrument 4908, Grant, Edvard Omholt-Jensen to Wilfred Hugo Doherty, Registered October 16, 1945, Muskoka Land Registry Office #35, Bracebridge, Ontario.

17 Odegard Omejer to Mrs. H.C.P. Hazlewood, Letter, July 10, 1947, Large Hol. Box 47B, Gravenhurst Archives, Gravenhurst, Ontario.

18 Norway, "Norway and Canada During World War 2," Royal Norwegian Embassy in Ottawa, http://www.norway.no/en/canada/norway-canada/historical-ties/#NorwayandCanadaduringWorldWar2 (Accessed March 19, 2017).

19 Looseley, "History of the Royal Norwegian Air Force in Canada," 19, 45, 98, 99.

20 Norway, "Norway and Canada During World War 2," Royal Norwegian Embassy in Ottawa, https://www.norway.no/en/canada/norway-canada/historical-ties/#NorwayandCanadaduringWorldWar2 (Accessed March 19, 2017).

21 "Norwegian Aces," Wall Text, Little Norway Memorial, Muskoka Airport, Gravenhurst, Ontario.

22 Wilhelm Mohr, "The Contribution of the Norwegian Air Forces," *Britain and Norway in the Second World War*, 92.

23 Ibid., 89.

24 Ibid., 93.

Epilogue

1 "Camp Norway Monument," War Monuments in Canada, https://www.cdli.ca/monuments/ns/norway.htm (Accessed April 7, 2017).

2 Zena Cherry, "Norwegian Veterans Return," *Globe and Mail*, September 17, 1976; "Prince Unveils Little Norway Monument," *Globe and Mail*, September 20, 1976.

3 *Program for Royal Norwegian Air Force Reunion, June 14 to June 22, 1985*, Pamphlet, Gravenhurst, 1985, Karen Greavette Collection.

4 Alexandra Radkewycz, "Norwegian King Dedicates Park," *Globe and Mail*, November 21, 1987; Zena Cherry, "Bank Presents Trophy for Equestrian Event," *Globe and Mail*, November 21, 1987.

5 Al Bacon, Personal Interview with Author, Bracebridge, Ontario, September 2, 2014; Kate Harries, "Muskoka Honours 'Little Norway,'" *Toronto Star*, May 6, 2002.

6 Tom Lee to Albert H. Bacon, Letter, October 13, 2000, Bacon Family Collection.

7 Allan Cook, "Saluting Little Norway's Legacy," *Muskoka Magazine*, November 2007, 64.

8 Lars Lindgren, Telephone Interviews with Author, January 9, 2015 and January 24, 2017. Lindgren is an aviation historian and past president of the 451 Norwegian Wing of the Royal Canadian Air Force Association.

9 Sheila Copps to Albert H. Bacon, Letter, u.d., Bacon Family Collection.

10 Brad Hammond, "As Time Goes By: The Vikings Return," *Vintage Muskoka*, May 2002, 13–15, File 05–45, Gravenhurst Archives, Gravenhurst, Ontario; Mark Clairmont, "Royal Occasion at Royal York," *Muskoka Today*, May 10, 2002.

11 Al Bacon, Personal Interview with Author, Bracebridge, Ontario, September 2, 2014.

12 Kate Harries, "Ceremony Honours 'Little Norway,'" *Toronto Star*, May 9, 2002; Allyson Snelling, "Royals Visit Muskoka," *Gravenhurst Banner*, May 10, 2002.

13 Boris Swidersky, "Funding Issues Continue for Project," *Examiner-Banner*, July 12, 2006.

14 Jacqueline Lawrence, "The Norwegian Memorial's Cost is Not Carved in Stone," *Examiner-Banner*, April 4, 2007.

15 Al Bacon, Personal Interview with Author, Bracebridge, Ontario, September 2, 2014.

16 Norway, "Norway and Canada During World War II," Royal Norwegian Embassy in Ottawa, https://www.norway.no/en/canada/norway-canada/historical-ties/#NorwayandCanadaduringWorldWar2; Per Olav Talgøy, "CNS Reports," *Canadian Nordic Society 2007 Newsletter*, November 2007, Gravenhurst Archives, Gravenhurst, Ontario.

17 Lars Lindgren, Telephone Interviews with Author, January 9, 2015 and January 24, 2017.

18 Inger Stene, Personal Interviews with Author, Toronto, Ontario, July 17, 2014, and August 27, 2014.

19 Neil Etienne, "Connection to Norway Honoured," *Gravenhurst Banner*, June 6, 2012.

Bibliography

Books

Andenæs, Johs., Olav Riste, Magne Skodvin. *Norway and the Second World War*. Oslo, Norway: Johan Grundt Tanum Forlag, 1966.

Andersson, Axel. *Hero for the Atomic Age: Thor Heyerdahl and the Kon-Tiki Expedition*. Oxford, United Kingdom: Peter Lang Ltd., 2010.

Citizens Committee for Recreation of Troops in Training. *Active Service Canteen Toronto 1939–1945*. Toronto, Ontario: Bryant Press Ltd., 1945.

Germany. *Documents on German Foreign Policy 1918–1945*. Series D. 13 Volumes. Washington, D.C.: U.S. Government Printing Office, 1954.

Germany: Kriegsmarine Oberkommando. *Fuehrer Conferences on Matters Dealing with the German Navy, 1940*. 7 Volumes. Washington, D.C.: U.S. Navy Department, 1947.

Gravenhurst Historical Committee. *The Light of Other Days*. 2nd ed. Cobourg, Ontario: Gravenhurst Historical Committee, 1972.

Guhnfeldt, Cato. *Spitfire Saga*. 5 Volumes. Oslo, Norway: Wings Forlag, 2009–2014.

Hansen, Per Conradi. *Little Norway: A Message of Liberty to the Hills of Home*. Oslo, Norway: The Military Historical Foundation of Eastern Norway, 1985.

Henriksen, Dag. "Biography of Wilhelm Mohr." *Wilhelm Mohr on World War II*, comp. by Dag Henriksen. Trondheim, Norway: Tapir Academic Press, 2009, 13–20.

Howarth, David. *The Shetland Bus*. Guilford, Connecticut: Lyons Press, 1951.

Hvidsten, J. Peter. "Per Hvidsten: An Amazing Journey, Norway to Canada." Biography. Port Perry, Ontario, 2012. J. Peter Hvidsten Collection.

International Military Tribunal. *Nazi Conspiracy and Aggression*. 8 Volumes. Washington, D.C.: U.S. Government Printing Office, 1945–1946.

International Military Tribunal. *Trial of the Major War Criminals before the International Military Tribunal, Nuremberg, 14 November 1945–1 October 1946*. 22 Volumes. Nuremberg, Germany: International Military Tribunal, 1947.

Jenkins, Roy. *Churchill: A Biography*. New York, New York: Penguin Group, 2002.

Jensen, Harald. *The Adrenalin Years*. Don Mills, Ontario: Frances M. Jensen, 1992.

Larsen, Tor Idar and Finn Thorsager. *Viking Spitfire: The Story of Finn Thorsager*. United Kingdom: Fonthill Media, 2012.

Looseley, H.W. "History of the Royal Norwegian Air Force in Canada." History. Toronto, 1945. L Box, Class 385.4 (481), Norwegian Armed Forces Museum Library, Oslo, Norway.

Lunde, Henrik O. *Hitler's Pre-Emptive War: The Battle for Nor-*

way, 1940. Havertown, Pennsylvania: Casemate, 2010.

Mazower, Mark. *Hitler's Empire: How the Nazis Ruled Europe*. London, England: Penguin Books, 2008.

McGuire, C.R. "The 'Little Norway' Story." History. Ottawa, 1980. Large Hol. Box 47B, Gravenhurst Archives, Gravenhurst, Ontario.

Mohr, Wilhelm. "9 April—From Past to the Future." *Wilhelm Mohr on World War II*, comp. by Dag Henriksen. Trondheim, Norway: Tapir Academic Press, 2009, 101–114.

Mohr, Wilhelm. "The Contribution of the Norwegian Air Forces." *Britain and Norway in the Second World War*, ed. by Patrick Salmon. London, England: H.M. Stationery Office, 1995, 83–96.

Mohr, Wilhelm. "The Contribution of the Norwegian Air Forces." *Wilhelm Mohr on World War II*, comp. by Dag Henriksen. Trondheim, Norway: Tapir Academic Press, 2009, 29-78.

Mohr, Wilhelm. "The Ground Crews' Participation in the War on the Continent 1944/45. " *Wilhelm Mohr on World War II*, comp. by Dag Henriksen. Trondheim, Norway: Tapir Academic Press, 2009, 79–93.

North, Harold. "Beyond the Seas." Manuscript. Gravenhurst, Ontario, n.d. Olafson Family Collection.

Norway. "The History of 'Little Norway' in Canada." History. Ottawa, 1987. PortsToronto Archives, Toronto, Ontario.

Olafson, Harald. "Little Norway in Muskoka." *Light of Other Days*. 2nd ed. Gravenhurst Historical Committee. Cobourg, Ontario: Gravenhurst Historical Committee, 1972, 124–126.

Pearce, M.J. and R. Porter. Introduction. *Fight for the Fjords: The Battle for Norway 1940*, ed. by Charlotte Carey. Plymouth, England: University of Plymouth Press, 2012, 1–15.

Pearson, Robert. *Gold Run: The Rescue of Norway's Gold Bullion from the Nazis, April 1940*. Oxford, England: Casemate, 2015.

Pedersen, Kai. *Viking Ships Had No Lifeboats and Other Stories*. Burk's Falls, Ontario: Arrow Printing and Publishing, 1995.

Porter, Cecil. *The Gilded Cage: Gravenhurst German Prisoner-of-War Camp 20, 1940–1946*. 2nd ed. Gravenhurst, Ontario: Gravenhurst Book Committee, 2003.

Reay Women's Institute. "Current Events (Dominion Airport)." "Tweedsmuir Village History." History. Gravenhurst Archives, Gravenhurst, Ontario, 37–39.

Reed-Olsen, Oluf. *Two Eggs on My Plate*. London, England: George Allen & Unwin Ltd., 1952.

Riste, Olav and Berit Nøkleby. *Norway 1940–1945: The Resistance Movement*. 3rd ed. Oslo, Norway: AIT Otta AS, 2004.

Saunders, S.J. Reginald. *Little Norway in Pictures: R.N.A.F. in Canada*. Toronto, Ontario: Norwegian Air Force in Exile, 1943.

Shirer, William L. *The Rise and Fall of the Third Reich: A History of Nazi Germany*. New York, New York: Simon & Schuster Paperbacks, 1959.

Singer, Kurt. *Duel for the Northland: The War of Enemy Agents in Scandinavia*. New York, New York: Robert M. McBride & Company, 1943.

Sønsteby, Gunnar. *Report from #24*. Fort Lee, New Jersey: Barricade Books Inc., 1999.

Stene, John Magne. "The Distinguished Flying Cross and How It Was Won." Autobiography. Toronto, Ontario, 1995. Inger Stene Collection.

Journals and periodicals

Guttman, Jon. "Norway's Ace of Aces." *Aviation History*, July 1999, 42–48.

Hansen, Per Conradi. "The Royal Norwegian Air Force in Canada, 1940–1945." *Canadian Defence Quarterly*, vol. 17, no. 2, Autumn 1987, 61–66.
Hatch, F. J. "The Spirit of Little Norway." *Canada's Wings: High Flight*, vol. 1, no. 5, April 1981, 171–180.
Talgøy, Per Olav. "CNS Reports." *Canadian Nordic Society 2007 Newsletter*, November 2007, Gravenhurst Archives, Gravenhurst, Ontario.

Parliamentary Documents

Canada

Canada. Parliament. House of Commons. *Debates*. 19th Parl., 1st sess., vol. 1, May 20, 1940.
Canada. Parliament. House of Commons. *Debates*. 19th Parl., 1st sess., vol. 1, May 27, 1940.

United Kingdom

HC Deb April 9 1940, vol 359, col 509
HC Deb May 2 1940, vol 360, col 907–910
HC Deb May 7 1940, vol 360, cols 1076, 1077
HC Deb May 8 1940, vol 360, cols 1353, 1354
HL Deb May 8 1940, vol 116, cols 296, 297

Newspapers

Bracebridge Gazette

"Aeroplane Crash at Germania." *Bracebridge Gazette*, August 31, 1944.
"Aeroplane Crash at Reay." *Bracebridge Gazette*, October 19, 1944.
"Army Train Visits Bracebridge." *Bracebridge Gazette*, February 12, 1942.
"County Town Celebrates Fall of Japan." *Bracebridge Gazette*, August 16, 1945.
"Monday's Celebration." *Bracebridge Gazette*, May 10, 1945.
"Serious Fire at Camp Little Norway." *Bracebridge Gazette*, February 23, 1945.

Evening Telegram (Toronto)

"Active Service Canteen is Well Patronized." *Evening Telegram*, January 29, 1941.
"Allies Mine Norse Waters Cripple All German Trade Hope to Bottle Nazi Fleet." *Evening Telegram*, April 8, 1940.
"Between You and Me." *Evening Telegram*, November 11, 1940.
"British Ship Goes Down Sinking Sub." *Evening Telegram*, April 8, 1940.
"Canada Airmen are Preferred as Instructors." *Evening Telegram*, May 7, 1940.
"Canadian and Norwegian Airmen Get Turkey Feast." *Evening Telegram*, December 24, 1940.
"Cheers Echo in Commons as Challenge of Labour 'Welcomed' by Premier." *Evening Telegram*, May 8, 1940.
"Cloud Combing: The Norwegians." *Evening Telegram*, February 18, 1941.
Cole, Percy T. "Fighters Technical Winners as Norse 'Bomb' Toronto but Island Port 'In Ruins.'" *Evening Telegram*, March 10, 1941.
"Coroner Fears Airplane May Fall in City Street if Flying Not Controlled." *Evening Telegram*, July 12, 1941.
"Crown Prince Finds Laughter in Camp Name." *Evening Telegram*, May 5, 1942.
"End of War in Norway Leaves Nazi Sea Route Open." *Evening Telegram*, June 10, 1940.

"Flyers from Little Norway Down 14 Nazis at Dieppe." *Evening Telegram*, August 28, 1942.

"Haakon Hounded by Nazis Refuses to Leave Country." *Evening Telegram*, April 13, 1940.

"Heavy German Warships Moved to Norse Coast Landing Forces Stand By." *Evening Telegram*, April 8, 1940.

"Holiday-Makers Pass Prison Camp on Luxurious Vessel with Band." *Evening Telegram*, August 17, 1940.

"Largest Parade in City Since First Great War Watched by 100,000." *Evening Telegram*, February 22, 1941.

"Nazis Feel Vengeance of Norwegian Airmen Who Graduated Here." *Evening Telegram*, August 7, 1942.

"Norse Airman Dies in Crash in Mock Fight." *Evening Telegram*, January 21, 1942.

"Norse Battle 'Quislings' in Bitter Street Clashes." *Evening Telegram*, February 5, 1941.

"Norse Bishops Openly Attack Nazi Overlords." *Evening Telegram*, February 4, 1941.

"Norse Camp Inaugurated as Flyers Continue Fight to Lift Yoke from Nation." *Evening Telegram*, November 11, 1940.

"Norse Flyers Do Air Stunts on Their Skis." *Evening Telegram*, January 13, 1941.

"Norse Royalty Safe After Long Voyage." *Evening Telegram*, August 29, 1940.

"Norse Victims of Air Smash Badly Injured." *Evening Telegram*, October 24, 1940.

"Norway Plans Training Pilots at Local Field." *Evening Telegram*, July 4, 1940.

"Norwegian Airmen Feature First Military Ski Meet." *Evening Telegram*, February 3, 1941.

"Norwegian Navy Flyers to Train at Vancouver." *Evening Telegram*, January 28, 1941.

"Norwegians Attend Supper." *Evening Telegram*, December 23, 1940.

"Permit Norwegian to Train Airmen." *Evening Telegram*, July 5, 1940.

"Planes at Island Airport Dip Their Wings in Salute to Royal Norwegian Couple." *Evening Telegram*, January 14, 1941.

"Planes Crash in Mid-Air One Killed, Two Unhurt." *Evening Telegram*, July 23, 1942.

"Powerful Units of Navy and Huge Allied Air Force Join Battle Beside Norse." *Evening Telegram*, April 9, 1940.

"Prince Harald of Norway Wins Hearts of Fighters Training Near Huntsville." *Evening Telegram*, September 23, 1942.

"Proud 'Little Norway' Camp Thrilled by Royal Visitors." *Evening Telegram*, January 13, 1941.

Pyper, C.B. "Brilliant Norway Action Shows British Better Men Than Nazis Ever Will Be." *Evening Telegram*, May 7, 1940.

"Report British Advancing to Major Clash with Foe in North Trondheim Zone." *Evening Telegram*, April 18, 1940.

"Royal Norwegian Couple Given Rousing Welcome by Gravenhurst Throng." *Evening Telegram*, May 5, 1942.

"Royal Visitors Attend Official Opening of New Norwegian Camp in Muskoka." *Evening Telegram*, May 4, 1942.

"Seasonable Word of Warning from the Chief Coroner." *Evening Telegram*, July 14, 1941.

"Snow-Clad Hills of North Delight Buffalo Skiers." *Evening Telegram*, January 20, 1941.

"Stunned Danes Awake to Find Armed Nazis Posted Throughout City." *Evening Telegram*, April 10, 1940.

"Toronto Bride of 2 Months Bereaved as Two Killed in Crash of Norse Plane." *Evening Telegram*, February 22, 1941.

Wuorio, Eva-Lis. "New Lodge in Ontario Woods Prepares Norse for War." *Evening Telegram*, January 31, 1942.

"Youthful Prince Harald of Norway Visits His Airmen and Proves Real 'Prince Charming' to Entire Camp." *Evening Telegram*, September 24, 1942.

Globe and Mail

"2 Norse Fliers Hurt in Crash." *Globe and Mail*, October 24, 1940.

"2 Norse Fliers Killed in West." *Globe and Mail*, March 19, 1941.

"4 Airmen Killed Here When 2 Bombers Crash." *Globe and Mail*, July 7, 1941.

"2-Day Hunt Fails to Find Norse Pilot and Plane." *Globe and Mail*, January 26, 1942.

"Anonymous Norsemen Play Great Part in War." *Globe and Mail*, March 24, 1942.

"Camp Little Norway is Sold; Becomes Civilian Airfield." *Globe and Mail*, August 23, 1945.

"Canadian Brides Happy in Norway, Says War Flier." *Globe and Mail*, June 21, 1946.

"Canadian Brides Have Housing Trouble in Oslo." *Globe and Mail*, June 11, 1946.

Cherry, Zena. "Bank Presents Trophy for Equestrian Event." *Globe and Mail*, November 21, 1987.

Cherry, Zena. "Norwegian Veterans Return." *Globe and Mail*, September 17, 1976.

"Crash Kills Norse Flier Who Wed Toronto Girl." *Globe and Mail*, February 22, 1941.

"Crash Victim's Wife, Son Free." *Globe and Mail*, February 24, 1943.

"Don't Want to Return as Instructors, Fighting Norse Declare Otherwise." *Globe and Mail*, August 7, 1942.

"Eyes Turn to Sky as Norse Planes Zoom." *Globe and Mail*, March 25, 1941.

"Fanny and Her Furry Family Will Shock Norwegian Flier." *Globe and Mail*, August 23, 1945.

"Ferry Fliers Outgrow Port at Montreal." *Globe and Mail*, July 27, 1942.

Frayne, Trent. "Grateful Forever to Canadians, Says Commander." *Globe and Mail*, February 17, 1945.

"Funeral Is Held for Two Fliers." *Globe and Mail*, May 2, 1941.

"Grace Notes." *Globe and Mail*, May 2, 1942.

"'Grads' of Little Norway Earn Praise Overseas." *Globe and Mail*, December 24, 1941.

"Little Bergen Dedicated by Norwegians." *Globe and Mail*, May 5, 1942.

"Little Norway Doors to Open." *Globe and Mail*, October 7, 1944.

"Little Norway Men Fought at Dieppe." *Globe and Mail*, August 27, 1942.

"Lyric Male Choir Host to Norwegian Singers." *Globe and Mail*, April 8, 1942.

MacTaggart, Ken W. "Bright Record Written by Airmen." *Globe and Mail*, March 9, 1943.

MacTaggart, Ken W. "Latest 'Little Norway' Gives Men Home Flavor." *Globe and Mail*, March 10, 1942.

"Mary Falla is Married to Lieut. H. Olafson." *Globe and Mail*, February 18, 1942.

"New Garb Styled at Little Norway." *Globe and Mail*, November 17, 1943.

"Norse Airman Listed Missing." *Globe and Mail*, January 10, 1942.

"Norse Airmen Awarded D.F.C." *Globe and Mail*, September 7, 1942.

"Norse Airmen Cleared of 5th Column Suspicion." *Globe and Mail*, July 16, 1941.

"Norse Airmen to Share Yule Gifts with Sailors." *Globe and*

Mail, December 25, 1941.
"Norse Flier, 23, Killed in Mid-Air Collision." *Globe and Mail*, July 23, 1942.
"Norse Flying Student to Marry Toronto Girl." *Globe and Mail*, January 4, 1941.
"Norse in Ontario Honor Prince Olav and Wife." *Globe and Mail*, January 20, 1942.
"Norse Prince Dedicates New R.N.A.F. School." *Globe and Mail*, May 5, 1942.
"Norway Flier Dies in Crash." *Globe and Mail*, February 23, 1943.
"Norway Raids Boost Morale, Prince Says." *Globe and Mail*, January 21, 1942.
"Norwegian Airman Dies in Crash Near Oakville." *Globe and Mail*, January 21, 1942.
"Norwegian Airman is Killed at Norval." *Globe and Mail*, January 25, 1943.
"Norwegian Airmen Ready for Invasion of Their Canadian Wives and Children." *Globe and Mail*, August 24, 1945.
"Norwegian OC Charges Fifth Columnists Trying to Undermine Efforts." *Globe and Mail*, May 13, 1941.
"Norwegians Say Farewell." *Globe and Mail*, February 19, 1945.
"Ola Adds to Ski Laurels." *Globe and Mail*, February 9, 1942.
"Ole Reistad Tops Field in Ski Test." *Globe and Mail*, March 9, 1942.
Palmer, Patricia. "Norway Has a 'Little Canada' Now, Many of the Girls from Toronto." *Globe and Mail*, December 17, 1946.
"Pilot Lands Safely but Plane Hits Tree." *Globe and Mail*, December 27, 1940.
"Plane Search Still Fruitless." *Globe and Mail*, January 27, 1942.
"Prince Unveils Little Norway Monument." *Globe and Mail*, September 20, 1976.
"Quit Search for Plane and Pilot, Lieut. T. Nalki." *Globe and Mail*, January 28, 1942.
Radkewycz, Alexandra. "Norwegian King Dedicates Park." *Globe and Mail*, November 21, 1987.
"Rain Hampers Norse Search." *Globe and Mail*, April 21, 1941.
Sanderson, Eric. "R.N.A.F. Members' Wives Learn About Life in Norway." *Globe and Mail*, August 19, 1943.
"Ski Troopers Feature of Huntsville Tourney." *Globe and Mail*, February 3, 1941.
Smith, Earl. "100,000 Line Streets to See Five-Mile Parade of Troops and Civilians Spur Savings Drive." *Globe and Mail*, February 24, 1941.
Smith, Earl. "Campaign Success Sure, Celebrate on Saturday." *Globe and Mail*, March 14, 1941.
Smith, Earl. "Little Norway is Inspected by Prince Olav." *Globe and Mail*, January 14, 1941.
Smith, Earl. "Norse Airmen Do Ski Jumps at Huntsville." *Globe and Mail*, January 13, 1941.
"Soccer Team to Chicago." *Globe and Mail*, June 12, 1942.
"Sonja Flies to Muskoka to Visit Norse Airmen." *Globe and Mail*, March 23, 1944.
"Stamp to Honor 'Little Norway,'" *Globe and Mail*, February 13, 1946.
"Stress Danger Novice Pilots Are to Citizens." *Globe and Mail*, July 12, 1941.
"Their Old Yule Customs are Revived in Toronto by Norse Fighting Men." *Globe and Mail*, December 25, 1940.
"Thousands on Beaches Watch Norwegian Flier Crash to Death in Lake." *Globe and Mail*, July 28, 1941.
"Toronto Little Norway Taken Over by R.C.A.F." *Globe and Mail*, April 6, 1943.
"Two Norwegian Fliers Unhurt in Plane Crash." *Globe and Mail*, December 27, 1940.

"Training Base of Norwegians Declared Open." *Globe and Mail*, November 11, 1940.

"Young Danish Flier Dies as Plane Dives into Lake." *Globe and Mail*, March 27, 1941.

Gravenhurst Banner, Gravenhurst News, Examiner-Banner

"Coca-Cola Co. Gives Stage Show for the R.N.A.F." *Gravenhurst Banner*, August 12, 1943.

"Defeat of Japan Again Brings Peace to War-Torn World." *Gravenhurst Banner*, August 16, 1945.

Etienne, Neil. "Connection to Norway Honored." *Gravenhurst Banner*, June 6, 2012.

"Farewell Ceremonies Held at Little Norway." *Gravenhurst Banner*, February 22, 1945.

"Gravenhurst Celebrates Defeat of Nazi Germany." *Gravenhurst Banner*, May 10, 1945.

Jardine, Albert. "Muskoka Airport, Part 3." *Gravenhurst News*, September 29, 1977.

Lawrence, Jacqueline. "The Norwegian Memorial's Cost is Not Carved in Stone." *Examiner-Banner*, April 4, 2007.

"Missing Prisoner Found in Camp." *Gravenhurst Banner*, April 8, 1943.

"Norwegian Royalty Visit Gravenhurst." *Gravenhurst Banner*, May 7, 1942.

"Officers' Mess Partly Destroyed." *Gravenhurst Banner*, February 22, 1945.

"Restaurant Closes on Wednesdays." *Gravenhurst Banner*, May 13, 1943.

"Salute to the Guard and Grand Parade Monday, May 24th." *Gravenhurst Banner*, May 20, 1943.

Snelling, Allyson. "Royals Visit Muskoka." *Gravenhurst Banner*, May 10, 2002.

"Summer Hotel for Norwegians." *Gravenhurst Banner*, August 21, 1941.

Swidersky, Boris. "Funding Issues Continue for Project." *Examiner-Banner*, July 12, 2006.

"Veterans Guard Honored Monday." *Gravenhurst Banner*, May 27, 1943.

"'Victory Parade' to Entertain Little Norway." *Gravenhurst Banner*, August 5, 1943.

"'Victory Train' Visits Gravenhurst." *Gravenhurst Banner*, February 12, 1942.

"W.H. Doherty Purchases Airport." *Gravenhurst Banner*, August 23, 1945.

Huntsville Forester, Forester

"Around Town." *Huntsville Forester*, July 31, 1941.

"Big Events Scheduled for Over Week-End." *Huntsville Forester*, January 30, 1941.

"Countryside News." *Huntsville Forester*, October 2, 1941.

"First Military Ski Meet is a Huge Success." *Huntsville Forester*, February 6, 1941.

"Gravenhurst Celebrates Victoria Day." *Huntsville Forester*, May 27, 1943.

"Huntsville Hysterically Happy." *Huntsville Forester*, August 16, 1945.

"Lieut.-Col. Ole Reistad Winner of Long Ski Race." *Huntsville Forester*, March 12, 1942.

"Little Norway Bids Farewell to Canada." *Huntsville Forester*, February 22, 1945.

"Little Norway Locates Camp at Emsdale." *Huntsville Forester*, July 10, 1941.

"Norwegians Move to Muskoka." *Huntsville Forester*, March 5, 1942.

Pedersen, Kai. "Little Norway May She Always Be Free." *Forester*, August 19, 1981.
Rice, Paul H. "Around Town." *Huntsville Forester*, January 9, 1941.

Muskoka Sun

McClellan, Scott. "Recalling the Spirit of Little Norway, Part I: The Escape." *Muskoka Sun*, June 6, 1985.
McClellan, Scott. "Recalling the Spirit of Little Norway, Part II: Toronto." *Muskoka Sun*, June 13, 1985.
McClellan, Scott. "Recalling the Spirit of Little Norway, Part III: Muskoka." *Muskoka Sun*, June 20, 1985.
Sparrow, Catherine Groth. "The Spirit of Little Norway–Part II." *Muskoka Sun*, June 25, 1981.
Stone, Dennis. "R.N.A.F. Bases in Muskoka During World War II." *Muskoka Sun*, August 23, 1973.

Muskoka Today

Clairmont, Mark. "Royal Occasion at Royal York." *Muskoka Today*, May 10, 2002.
Porter, Cecil. "On the Homefront." *Muskoka Today*, May 7, 1995.

Toronto Daily Star, *Toronto Star*

"3 Norwegian Fliers are Awarded D.F.C." *Toronto Daily Star*, September 5, 1942.
"8 Months of Living Death Can't Break Norse Exiles." *Toronto Daily Star*, May 12, 1943.
"125 Viking War Aces to Roam Sky's Lanes Over Island Airport." *Toronto Daily Star*, July 4, 1940.
"A.A. Guns on Way to Norway Sunk by Sub, Hankey Says." *Toronto Daily Star*, May 8, 1940.
Aaron, Robert. "Hoard of Gold Coins Hidden During War Returned to Norway." *Toronto Star*, January 23, 1988.
"Airmen Get Turkey Feed at Howard Park Church." *Toronto Daily Star*, December 24, 1940.
"Allies May Invade 'Any Time Now,' Prince Says." *Toronto Daily Star*, January 20, 1942.
"Allies Mine Norse Waters Glad to Have Nazis Object." *Toronto Daily Star*, April 8, 1940.
"Bay Boats Ground to Bits Like Toys in Ripping Gale." *Toronto Daily Star*, September 26, 1941.
"Blond Babies Soon May See Land of Daddy." *Toronto Daily Star*, January 11, 1945.
"Buchenwald Victim to be Stalin Guest." *Toronto Daily Star*, June 4, 1945.
"Canadian Brides Off for Norway." *Toronto Daily Star*, September 24, 1945.
"Canuck Pilot Promises *Altmark* Picture Feast." *Toronto Daily Star*, March 27, 1940.
"Camp Stops Classes as Tribute to Flier." *Toronto Daily Star*, April 1, 1941.
"Carnival of Thrills and Spills as Army Ski Tourney Starts." *Toronto Daily Star*, February 1, 1941.
"Claims Norse Here 5th Column Target." *Toronto Daily Star*, May 13, 1941.
"Commanding Officer Leads His Men Home." *Toronto Daily Star*, March 9, 1942.
"Could Help Russ Problem by Recapturing Norway." *Toronto Daily Star*, August 13, 1942.
"Deaths: Selmer." *Toronto Daily Star*, January 15, 1943.
"Deaths." *Toronto Daily Star*, October 3, 1942.
"Dynamite Artist in Norway Builds Norse Air Base Here." *Toronto Daily Star*, March 4, 1942.
"Escaped from Nazis, Norwegian Weds Toronto Girl." *Toronto*

Daily Star, December 27, 1940.
"Ever Grateful Declare Norse Leaving Canada." *Toronto Daily Star*, February 17, 1945.
"Every Enemy Warship in Norwegian Waters Bottled Up by Allies." *Toronto Daily Star*, April 10, 1940.
"Fifty Buffalo Skiers Declare First 'International Ski Week-End' at Huntsville Great Success." *Toronto Daily Star*, January 20, 1941.
"First Wren from Norway Longs to be Back on Skis." *Toronto Daily Star*, March 15, 1944.
"Fled Norway, Trained Here, Missing After Air Fight." *Toronto Daily Star*, January 9, 1942.
"Flier Still Unconscious in First Norse Air Crash." *Toronto Daily Star*, October 24, 1940.
"Flying to Dead Wife, Norwegian Airman Crashes." *Toronto Daily Star*, December 27, 1940.
"'For Norway,' Is War Cry of Viking Air Force Girls." *Toronto Daily Star*, November 16, 1943.
"Girls Learn Norse Tongue So Can Talk to Hubbies." *Toronto Daily Star*, July 22, 1943.
"God Send Us Home in Peace Norse Prince, Airmen Pray." *Toronto Daily Star*, January 14, 1941.
"Gold Out by Back Door as Nazi Entered Front," *Toronto Daily Star*, January 22, 1941.
Harries, Kate. "Ceremony Honors 'Little Norway.'" *Toronto Star*, May 9, 2002.
Harries, Kate. "Muskoka Honors 'Little Norway.'" *Toronto Star*, May 6, 2002.
"Have Danced '1,000 Miles' with Allied Servicemen." *Toronto Daily Star*, January 24, 1944.
"'Haven't Taken Off Boots Since Tuesday–Haakon 'Civilization At An End.'" *Toronto Daily Star*, April 13, 1940.
"Havergal Student Guest of Royalty." *Toronto Daily Star*, January 20, 1942.
"Helpless Norse Sailors Machine-Gunned in Water." *Toronto Daily Star*, April 24, 1940.
"Herd 500 Norse Teachers in Ship Crowded by 250." *Toronto Daily Star*, April 27, 1942.
"Here Comes Ola in Action Shots Showing Ski-Jumping Made Easy." *Toronto Daily Star*, February 14, 1942.
"Here to Pick Up Warplanes 100 of Them for Norway." *Toronto Daily Star*, April 26, 1940.
"Hills, Bush Recall Native Land to Norse Training at Emsdale." *Toronto Daily Star*, October 11, 1941.
"Honor Two Choirs." *Toronto Daily Star*, June 13, 1942.
"Hope Ebbs Pilot XXX Will Be Found Alive." *Toronto Daily Star*, April 26, 1941.
"International Skiers on Sports Week-End." *Toronto Daily Star*, January 18, 1941.
Jones, Donald. "Unravelling 'Little Norway's Big Secrets." *Toronto Star*, May 10, 1986.
"Lack of 300 Feet Spelt Death for Norse in 400 m.p.h. Dive." *Toronto Daily Star*, January 22, 1942.
"Land Plane in Field 'Could Have Crashed.'" *Toronto Daily Star*, February 12, 1942.
"Little Norway Air School Turned Over to R.C.A.F." *Toronto Daily Star*, April 6, 1943.
"Little Norway Does Big Share in First Year." *Toronto Daily Star*, November 11, 1941.
"Little Norway's Graduates Busy Hunting Nazis." *Toronto Daily Star*, July 16, 1942.
"Mary Frances Loosemore Norwegian Airman's Bride." *Toronto Daily Star*, December 26, 1940.
"Men Went Mad on *Altmark* Says Man Who Was There."

Toronto Daily Star, June 6, 1940.
"Men's Choirs Co-operate." *Toronto Daily Star*, May 2, 1942.
Menken, Arthur. "Norwegian College Boys Thumb Way Home to War." *Toronto Daily Star*, April 15, 1940.
Milsom, Patrick. "Norse 'Ulysses' Outwits Nazis–Now He Can Fight Back." *Toronto Daily Star*, November 15, 1940.
Morton, Paul. "Norse Fliers Sure Now That 5-Year-Old Prince is True Son of Vikings." *Toronto Daily Star*, September 23, 1942.
"Must Find Norse Flier in 3 Days or 'He's Dead.'" *Toronto Daily Star*, April 25, 1941.
Myre, Olav. Untitled. *Toronto Daily Star*, April 9, 1940.
"Nazi Bombs Rain All Night Swedes Fear Next Thrust." *Toronto Daily Star*, April 12, 1940.
"Nazi Bullies on *Altmark* Kicked 13 Year-Old Lad." *Toronto Daily Star*, September 10, 1941.
"Nazis Sending Families Home Norse Situation 'Near Explosion.'" *Toronto Daily Star*, October 17, 1944.
"New Norwegian Air Camp Is Opened at Gravenhurst." *Toronto Daily Star*, May 4, 1942.
"Norse Airman Now Convinced That Sign Meant What It Said." *Toronto Daily Star*, September 28, 1940.
"Norse Flier Falls in Lake Before Eyes of Thousands." *Toronto Daily Star*, July 28, 1941.
"Norse Flier, Plane Sought by Nearly 30 War Craft." *Toronto Daily Star*, April 24, 1941.
"Norse Fliers Swim, Walk, Row to Rejoin Legion Here." *Toronto Daily Star*, August 6, 1940
"Norse Fliers Tested in 20-Mile Ski Trek." *Toronto Daily Star*, March 9, 1942.
"Norse Fliers Win R.A.F. Praise for Daredevil Daylight Raids." *Toronto Daily Star*, December 23, 1941.
"Norse Giving In, Berlin Says, but Only Inch at a Time." *Toronto Daily Star*, April 10, 1940.
"Norse Naval Fliers to Train at Coast." *Toronto Daily Star*, January 28, 1941.
"Norse Patriots Dare Death So People Can Get News." *Toronto Daily Star*, January 5, 1942.
"Norse Plane in Formation Spins to Earth, Pilot Dies." *Toronto Daily Star*, January 21, 1942
"Norse Women's Auxiliary Prepares for Liberation." *Toronto Daily Star*, May 16, 1944.
"Norsemen Greet with Joy Soviet Liberating Forces." *Toronto Daily Star*, October 27, 1944.
"Norsemen Hurtle to Death Amid Fiery Plane Wreckage." *Toronto Daily Star*, July 5, 1941.
"Norsemen Invade Homeland to Help Russians Oust Foe." *Toronto Daily Star*, November 14, 1944.
"Norway Flier in Crash Fog Hid Rest of Flight." *Toronto Daily Star*, January 25, 1943.
"Norway Fliers Trained Here Now in Hot Pursuit of Nazis." *Toronto Daily Star*, August 7, 1942
"Norwegian Air Force Trainees Join Hunt for Escaped Nazi." *Toronto Daily Star*, December 10, 1942.
"Norwegian Airmen-Singers to Join Coliseum Chorus." *Toronto Daily Star*, April 14, 1942.
"Norwegian Army Airmen Train at Island Airport to Strike Another Blow for Homeland." *Toronto Daily Star*, October 24, 1940.
"Norwegian Auxiliary Will Hold Carnival." *Toronto Daily Star*, October 2, 1944.
"Norwegian Crews Drove Trains Right Up to Front to Get

Them Returning British Soldiers Say." *Toronto Daily Star*, May 7, 1940.
"Norwegian Fliers Clean Up in Huntsville Ski Carnival." *Toronto Daily Star*, February 3, 1941.
"Norwegian Guests of Kiwanis Women." *Toronto Daily Star*, December 23, 1940.
"Norwegian Hero." *Toronto Daily Star*, January 28, 1942.
"Norwegian Keeps Date and Nearly Misses Baton!" *Toronto Daily Star*, June 5, 1942.
"Norwegian Pilots to Train Here for Further Combat." *Toronto Daily Star*, July 20, 1940.
"Norwegian Royalty Skiing in Huntsville." *Toronto Daily Star*, January 19, 1942.
"Norwegian Skier Wins U.S. Title." *Toronto Daily Star*, February 9, 1942.
"Norwegian Trainees Moving to Muskoka." *Toronto Daily Star*, February 21, 1942.
"The Norwegians Were Once Great Whale Fishers." *Toronto Daily Star*, January 29, 1942.
"Off to Invade England Nazi Idea Till Oslo Near." *Toronto Daily Star*, April 18, 1940.
"Oil Patch on Lake May Hide Plane." *Toronto Daily Star*, April 19, 1941.
"Olsen, Nils," Lieutenant. "Norse Fugitives' Ship Hit Downed Two Raider Planes." *Toronto Daily Star*, November 13, 1941.
"Only 18 Norse Clergy Quisling 1,500 Quit Jobs." *Toronto Daily Star*, March 4, 1942.
"Oslo Folk Flee in Droves Fearing British-Nazi Fight." *Toronto Daily Star*, April 11, 1940.
Ouzounian, Richard. "Scenes from a Life Lived Large." *Toronto Star*, March 14, 2015.
"Over the Teacups." *Toronto Daily Star*, October 3, 1940.
"Over the Teacups." *Toronto Daily Star*, November 11, 1940.
"Over the Teacups." *Toronto Daily Star*, December 18, 1940.
"Over the Teacups." *Toronto Daily Star*, February 13, 1942.
"Picked 7 War Victors U.S. Flier Joins Norse." *Toronto Daily Star*, July 7, 1941.
"Planes Collide in Mid-Air Norwegian Flier is Killed." *Toronto Daily Star*, July 23, 1942.
Plewman, W.R. "The War Reviewed." *Toronto Daily Star*, April 8, 1940.
"Quisling Rules Churches Only 30 of 1,000 Are Open." *Toronto Daily Star*, April 14, 1942.
"Reprisals Follow Raid on Islands Off Norway." *Toronto Daily Star*, March 7, 1941.
Rhodes, Peter C. "2,000 Germans Kill 542 Norse to Take Narvik." *Toronto Daily Star*, April 10, 1940.
"Roars at 300 M.P.H. to Earth and Death." *Toronto Daily Star*, February 23, 1943.
"Rough, Tough Foes of Nazis are Norse Commando Dogs." *Toronto Daily Star*, February 11, 1943.
"Sabotage Cheered by Norse Princess." *Toronto Daily Star*, September 27, 1941.
"Saw 'Chute Drop from Sky 'Must Have Gone Into Lake.'" *Toronto Daily Star*, January 26, 1942.
"See Norse Fliers Swimming after Eastern Gap Crash." *Toronto Daily Star*, March 26, 1941.
"Ship Which Survived Two Wars Will House Norse Fliers Here." *Toronto Daily Star*, September 6, 1940.
"Ski Train and Ski Trail Cement Anti-Axis Bonds." *Toronto Daily Star*, January 20, 1941.
"Ski Troops from Many Lands Compete at Huntsville Meet."

Toronto Daily Star, February 3, 1941.

"Society Highlights: Toronto's Norse Residents Greet Prince and Princess." *Toronto Daily Star*, January 14, 1941.

"Soldier for Dinner Plan Over the Top in Toronto." *Toronto Daily Star*, December 27, 1940.

"Soldiers and New Year Parties." *Toronto Daily Star*, January 2, 1941.

"Soldiers Seen Leaping Off Ship that Ignored Sub's Orders to Stop." *Toronto Daily Star*, April 8, 1940.

"Sonja Gets Rubber Boots to Tour Little Norway." *Toronto Daily Star*, March 23, 1944.

"Southern Norsemen Send Plane as Gift." *Toronto Daily Star*, March 28, 1941.

"Stamp Corner." *Toronto Daily Star*, January 23, 1943.

"Studies Norse Year Says She's on Way 'Home'." *Toronto Daily Star*, October 27, 1945.

"Summer Hotel Deep in Bush Serves Norse Fliers Only." *Toronto Daily Star*, July 12, 1941.

"Ten Norwegian War Heroes Fly from Toronto to White House." *Toronto Daily Star*, September 25, 1941.

"Ten Norwegians Leave Toronto for Good-Will Tour of U.S." *Toronto Daily Star*, September 25, 1941.

"Think Dazzling Sunlight Caused Norse Fliers' Crash." *Toronto Daily Star*, June 21, 1941.

"Unconscious Eight Days." *Toronto Daily Star*, October 31, 1940.

"Vikings' Flag Flies Bravely Norse Warplanes Overhead." *Toronto Daily Star*, November 11, 1940.

"Watched R.A.F. Bombers Sink Nazi Cruiser." *Toronto Daily Star*, May 3, 1940.

"'We Hit Germans' Ship Explosion Was Terrific.'" *Toronto Daily Star*, April 12, 1940.

"'We're Coming Back Again' is Refrain as Norwegians Break Camp, Say Farewell." *Toronto Daily Star*, February 17, 1945.

"Willkie Carries Toronto Thousands Roar Acclaim." *Toronto Daily Star*, March 24, 1941.

"Willkie Laments Death of Norse Airman in Crash." *Toronto Daily Star*, March 27, 1941.

"With Canada's Airmen." *Toronto Daily Star*, October 15, 1940.

"Would Not Happen Again In 100 Years–Norseman." *Toronto Daily Star*, July 12, 1941.

Other Newspapers

Brown, Dave. "The Faithful Cat That Saved a Stranded Crew." *Ottawa Citizen*, October 8, 1999.

Carstensen, Svend. "Copenhagen Taken." *New York Times*, April 9, 1940.

"Goebbels Explains Move." *New York Times*, April 9, 1940.

"Reich Ship is Sunk." *New York Times*, April 9, 1940.

Schultz, Sigrid. "The Spirit of Little Norway." *Chicago Daily Tribune*, August 16, 1942.

Tebby, Mary-Lyn. "Vesle Skaugum: Oxbow Lake's Norwegian Past." *Muskokan*, August 5, 1999.

Thomas, Robert McG. Jr. "Max Manus, 81, Dies: Fought Nazis in Occupied Norway." *New York Times*, September 22, 1996.

Magazines

Cook, Allan. "Saluting Little Norway's Legacy." *Muskoka Magazine*, November 2007, 64–68.

Gold, L. "Know Your Hospital." *The Sanitarium Sun*, February 1954, 13.

Gumbert, Penny. "Muskoka Hospitable to Norwegian Servicemen." *Muskoka Magazine*, October/November 2005, 55–60.

Hammond, Brad. "As Time Goes By: The Vikings Return." *Vintage Muskoka*, May 2002, 13–15. File 05-45, Gravenhurst Archives, Gravenhurst, Ontario.

Hammond, Brad. "In Their Own Words: The Vikings in Muskoka." *Vintage Muskoka*, May 2002, 6, 7, 16, 17. File 05-45, Gravenhurst Archives, Gravenhurst, Ontario.

Hammond, Brad. "Room for the Sky and All Its Stars: Muskoka's Thor Heyerdahl." *Vintage Muskoka*, approx. 2001, 26–31.

McLaughlin, Paul. "The War Years." *Cottage Life*, May/June 1991, 57–60, 98, 100, 101.

Sandburg, Carl. "Carl Sandburg Says." *Wings for Norway*, 1942, 25, Karen Greavette Collection.

Tegen, Gunhild. "Little Norway." *Wings for Norway*, 1942, 14-19, Karen Greavette Collection.

"Under the Midnight Sun..." *Vintage Muskoka*, May 2002, 8, 9. File 05-45, Gravenhurst Archives, Gravenhurst, Ontario.

Pamphlets, Brochures, Cards

The Drama of Heavy Water. Brochure. Norway, n.d. Bacon Family Collection.

Norway. *HM King Olav V of Norway*. Pamphlet. N.d. Rita Hiorth Collection.

Program for Royal Norwegian Air Force Reunion, June 14–June 22, 1985. Pamphlet. Gravenhurst, 1985. Karen Greavette Collection.

Spirit of 'Little Norway.' Card. Toronto, n.d. Large Hol. Box 24, Gravenhurst Archives, Gravenhurst, Ontario.

Plaques, Wall Texts

Plaque. *The Royal Norwegian Air Force: 1940–1945*. Toronto Historical Board, 1984. Toronto, Ontario.

Wall Text. *Commonwealth Air Training Plan*. Canadian Warplane Heritage Museum. Hamilton, Ontario.

Wall Text. *Hitler Drops Quisling*. Norway's Resistance Museum. Oslo, Norway.

Wall Text. *Invasion of Norway, April 9, 1940*. Little Norway Memorial, 2007. Muskoka Airport, Gravenhurst, Ontario.

Wall Text. *Little Norway*. Little Norway Memorial, 2007. Muskoka Airport, Gravenhurst, Ontario.

Wall Text. *Norwegian Aces*. Little Norway Memorial, 2007. Muskoka Airport, Gravenhurst, Ontario.

Letters, Correspondence

Christensen, Nils. Letter. N.d. Little Norway Memorial. Muskoka Airport, Gravenhurst, Ontario.

Clement, Svend to Editor of *Gravenhurst Banner*. Letter. N.d.

Copps, Sheila to Albert H. Bacon. Letter. U.d. Bacon Family Collection.

Lee, Tom to Albert H. Bacon. Letter. October 13, 2000. Bacon Familly Collection.

Omejer, Odegard to Mrs. H.C.P. Hazlewood. Letter. July 10, 1947. Large Hol. Box 47B, Gravenhurst Archives, Gravenhurst, Ontario.

Royal Norwegian Air Force, Little Norway fonds. PortsToronto Archives, Toronto, Ontario.

Website Information

NOTE *End-of-line hyphens in this section are part of the website address.*

Agderposten. "Honored with a Memorial." Agderposten.no, http://www.agderposten.no/kjop-tilgang?aId=1.535919 (Accessed March 13, 2017).

BBC News. "Norway Apologizes for Deporting Jews During Holocaust." BBC News Services, http://www.bbc.com/news/world-europe-16761558 (Accessed March 16, 2017).

Briggs, Harold J. and Jane Briggs Van Buskirk. "The Mica Mine & Me." Fairy Lake Association, http://www.fairylakeassociation.com/history-stories-past-present/the-mica-mine-me/ (Accessed March 14, 2017).

"Camp Norway Monument." War Monuments in Canada, https://www.cdli.ca/monuments/ns/norway.htm (Accessed April 7, 2017).

Canada. "The British Commonwealth Air Training Plan." Veterans Affairs Canada, http://www.veterans.gc.ca/eng/remembrance/history/second-world-war/british-commonwealth-air-training-plan (Accessed March 12, 2017).

Canada. "D-Day and the Battle of Normandy." Veterans Affairs Canada, http://www.veterans.gc.ca/eng/remembrance/history/second-world-war/d-day (Accessed March 18, 2017)

Danish WW2 Pilots. "Bourbon de Parma, Prince Jacques de (1922–1964)." Danish WW2 Pilots, http://www.danishww2pilots.dk/profiles.php?id=248 (Accessed March 16, 2017).

Forsvarets Museer. "9 April 1940." Forvaretsmuseer.no, http://forsvaretsmuseer.no/Oscarsborg/Festningens-historie/9.april-1940 (Accessed March 7, 2017).

Gudbrandsdølen Dagningen. "Rolf Kolling (92) Met With President Obama." GD, https://www.gd.no/nyheter/rolf-kolling-92-motte-president-obama/s/1-934610-7404817 (Accessed March 18, 2017).

Gustavsson, Håkan. "Biplane Fighter Pilots Norway: Major Finn Thorsager." Håkan's Aviation Page, http://surfcity.kund.dalnet.se/norway_thorsagen.htm (Accessed March 7, 2017).

Gustavsson, Håkan. "Biplane Fighter Pilots Norway: Sergeant Per Waaler." Håkan's Aviation Page, http://surfcity.kund.dalnet.se/norway_waaler.htm (Accessed March 7, 2017).

Johnston, Timothy Allan. "Diary: No. 1 Service Flying Training School, Camp Borden." www.timothyallanjohnston.com, http://www.timothyallanjohnston.com/ (Accessed March 26, 2017).

Juno Beach Centre. "British Commonwealth Air Training Plan." Junobeach.org, http://www.junobeach.org/canada-in-wwii/articles/british-commonwealth-air-training-plan/ (Accessed March 12, 2017).

Larsson, Knut. "RAF Nordic Casualties of Crews and Losses of Aircraft in World War 2." Luftwaffe.no, http://www.luftwaffe.no/Table2.htm (Accessed March 9, 2017).

Muskoka Airport. "History." Muskoka Airport, http://www.muskokaairport.com/ (Accessed March 16, 2017).

Norges Bank. "Historical Exchange Rate Data, 1819–2003." norges-bank.no, http://www.norges-bank.no/en/Statistics/Historical-monetary-statistics/Historical-exchange-rates/ (Accessed March 9, 2017).

Norway. "Norway and Canada During World War 2." Royal Norwegian Embassy in Ottawa, https://www.norway.no/

en/canada/norway-canada/historical-ties/ #NorwayandCanadaduringWorldWar2 (Accessed March 19, 2017).

Pearson, Bob. "Max Manus–An Extraordinary Man." Norway During World War 2, http://www.nuav.net/maxmanus.html (Accessed March 16, 2017).

The Royal House of Norway. "Norway Thanks You." RoyalCourt.no, http://www.royalcourt.no/nyhet.html?tid=123132&sek=27262 (Accessed March 18, 2017).

Ships of the East Atlantic Line on Postcards. *Europa*. Mariners-l.co.uk, http://www.mariners-l.co.uk/eac/eac1930-1939.htm (Accessed March 10, 2017).

Spitfirepilots. "Per Waaler–Gloster Gladiator Pilot." Spitfirepilots.com, http://www.spitfirepilots.com/per-waaler-gloster-gladiator-pilot/ (Accessed March 16, 2017).

United States Holocaust Memorial Museum. "Norway." https://www.ushmm.org/wlc/en/articlephp?ModuleId=10005460 (Accessed March 16, 2017).

Ward, Geoff. "The Tempsford Taxis." Norway: War, Resistance, Peace, http://www.wwiinorge.com/our-stories/the-tempsford-taxis/ (Accessed February 7, 2017).

Warsailors. Warsailors.com, http://www.warsailors.com/freefleet/index.html (Accessed May 8, 2017).

World War II Today. "263 Squadron Land at Lake Lesjaskog." Ww2today.com, http://ww2today.com/24th-april-1940-263-squadron-land-at-lake-lesjaskog (Accessed March 9, 2017).

World War II Today. "Norwegian Shore Batteries Sink the Blucher." Ww2today.com, http://ww2today.com/norwegian-shore-batteries-sink-the-blucher (Accessed March 7, 2017).

Other

"Catalina-D-FP 183 Vingtor II." Article. Little Norway Memorial, Muskoka Airport, Gravenhurst, Ontario.

"Norwegian Air Force Camp to be Moved to Muskoka." Article. Anonby Family Collection.

"Norwegians Complete Long March–Camp Over Night at Lansing." Article. Anonby Family Collection.

"Obituary: Per Kaare Nilsen." Gravenhurst Archives, Gravenhurst, Ontario.

Odd Toldnes. Audio Tape of interview with Gravenhurst Archives, in "World War II Interviews: Muskoka Pioneers." 1989. Tape 97-18, Gravenhurst Archives, Gravenhurst, Ontario.

Property title information: Muskoka Land Registry Office #35, 15 Dominion Street, Bracebridge, Ontario. Land Registry Office #66, 420-20 Dundas Street West, Toronto, Ontario.

Interviews

Personal, Telephone and E-Mail Interviews Conducted by Author, 2014–2017

Laura Anonby
Lauretta Anonby
Larry Baadsvik
Al Bacon
Shirley Barlow
Alice Binder
Barbara Chramer
John Clement
Susan Whiteside Davis

Yvonne Dixon
Christine Cator Everest
Per Falkenhall Jr.
Knut Fossum
Cyril Fry
Marion Fry
Karen Greavette
Erik Hansen
Rolf Kristian Hauge
Rita Hiorth
Betty Honnor
Elaine Hooper
Sandra Howes
Randi Hvatum
J. Peter Hvidsten
Douglas Johnston
Yvonne Jones
Rolf Kolling
Kaye Kruusmagi
Earle Laycock
Lars Lindgren
Bonnie Lund
Nancy Lynch
Madeline Olsen
Patricia Pedersen
Irene Peters
Gunhild Ramm Reistad
Kjell Arne Reistad
Michael Skretteberg
Inger Stene
Ron Stoddart
Bill Whiteside

Index

Student pilots studying map, Little Norway, Muskoka Airport.

PHOTO: ANONBY FAMILY COLLECTION.